Design Engineer's Handbook

Design Engineer's Handbook

Keith L. Richards

CRC Press
Taylor & Francis Group
Boca Raton   London   New York

CRC Press is an imprint of the
Taylor & Francis Group, an **informa** business

CRC Press
Taylor & Francis Group
6000 Broken Sound Parkway NW, Suite 300
Boca Raton, FL 33487-2742

First issued in paperback 2017

No claim to original U.S. Government works
Version Date: 20120827

ISBN 13: 978-1-138-07694-5 (pbk)
ISBN 13: 978-1-4398-9275-6 (hbk)

Visit the Taylor & Francis Web site at
http://www.taylorandfrancis.com

and the CRC Press Web site at
http://www.crcpress.com

Contents

List of Figures

List of Tables

Preface

When writing a book on this subject, it is difficult to decide what to leave out. The world of mechanical design engineering is very broad and covers a wide range of subjects. This book is specifically aimed at the student design engineer who has left full- or part-time academic studies and requires a handy reference handbook. Some of the titles may seem a little obscure, but in my experience, working in a wide range of industries from machine tools to aerospace, I have used these subjects regularly.

The chapters on beams and torsion are included for obvious reasons. Why a chapter on limits and fits? Some student engineers have difficulty in determining the correct type of fit to specify and select either a combination that is too loose or one that is expensive to attain.

The chapter on lugs and shear pins is important in the design engineer's armory, as you will meet this design feature time and time again. It is quite surprising that some lug designs are either under- or overdesigned and have failed with very expensive results.

I have addressed issues with mechanical fasteners, more specifically bolts and screws. I have not discussed other forms of fasteners such as rivets, and so forth, because there was insufficient space to do the subject full justice. This may be left to future publications.

Because thick-wall or compound cylinders are not the exclusive province of the hydraulic or pneumatic engineer and are used quite extensively in mechanical engineering as connection features, I felt the subject should be covered.

The chapter covering helical compression springs will be useful to the student engineer. I debated whether to include tension and torsion springs but decided against that for the time being. This may be included in future publications. The helical compression spring is the most commonly used spring, which justified coverage in its own chapter.

I introduce the subject of analytical stress analysis using the Mohr's circle. It is surprising that there are so many student engineers who have only had a passing reference to the circle; it has been used by me extensively in my career, as it helps to explain quite complex stress issues. Two chapters are devoted to stress analysis covering both analytical and experimental analysis. Although the chapter on experimental stress analysis may be considered rudimentary, it will give the reader an introduction to the subject.

Fatigue and fracture have become very important subjects in engineering. At times, fatigue and fracture can produce a number of catastrophic failures, ranging from the early comet disasters to a number of bridges and walkways failing. In most cases, a premature failure due to fatigue can be attributed to design errors, and the design engineer must be on his guard when designing components that are subject to cyclic loading.

The chapter on gear systems should be self-explanatory. Sooner or later, the engineer will be faced with a problem regarding gearing, and some of the examples shown will be of assistance.

Finally, the chapter on cams and followers I consider important because the information here will be useful in a motion control problem; this will give the reader an insight into the issues associated with cams and followers.

1 Beams

1.1 BASIC THEORY

1.1.1 INTRODUCTION

The following are the minimum requirements of a straight beam:

- The beam is straight.
- The cross section is uniform along its length (L).
- The beam has at least one longitudinal plane of symmetry.
- The beam is long in proportion to its depth (d).
  - L/d is greater than 8 for metals with compact sections.
  - L/d is greater than 15 for beams with thin webs.
- All loads and reactions lie in the longitudinal plane of symmetry.
- The beam is not disproportionally wide.
- The maximum stress in the beam does not exceed the proportional limit (0.2%).

1.1.2 SIMPLE ELASTIC BENDING

When a beam is subjected to a transverse load, either a point load or an evenly distributed load along its length, a bending stress will exist within the beam cross section.

Considering the cantilever of a uniform section, as shown in Figure 1.1, subject to a bending moment, the bending stress will be zero at the neutral axis. The stress in the upper surface is in tension and the lower surface is in compression when subject to the bending moment.

In contrast, the stresses in the beam shown in Figure 1.2 are inverted in the beam, i.e., in compression in the upper surface and in tension in the lower surface. Within the elastic range of the material (i.e., below the elastic limit or the yield point) the bending stress (f_b) at any point in the cross-section (A:A) is

$$f_b = \frac{My}{I} \tag{1.1}$$

where
- M = bending moment at the section in question, Nm
- I = moment of inertia of the section m^4
- y = distance from the neutral axis to the point at which the stress is required
- f_b = bending stress (may be either tension or compression) (N/m^2)

Table 1.1 gives a selection of standard bending cases.

The bending moment M may be determined from the standard beam diagrams as shown in Table 1.1, where a list of several of these are shown alongside with formulas for deflection and shear.

Generally there is no interest in knowing the bending stresses within the beam. Usually the bending stress at the outer fiber is needed, as this is the maximum value. In unsymmetrical sections, the distance y must be taken in the correct direction across that portion of the section that is in tension

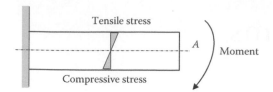

FIGURE 1.1 Stresses in a cantilever beam.

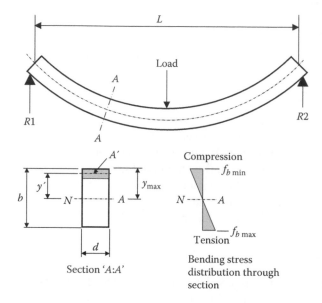

FIGURE 1.2 Stresses generated in a beam between supports.

or that portion in compression. Ordinarily only the maximum stress is needed, and this is the stress at the outer fiber under tension, which is at the greater distance y from the neutral axis.

1.1.3 Shearing Force and Bending Moment

The *shearing force* at a section of a beam is the algebraic sum of all the forces to one side of the section.

The *bending moment* at a section of a beam is the algebraic sum of the moments of all the forces to one side of the section.

At the point P in the cantilever shown in the Figure 1.3(a) the shearing force is $W_1 + W_2$, which is tending to shear the beam, as shown in Figure 1.3(b). This is opposed by the shearing resistance of the beam to the left of P.

The bending moment at the point P is $W_1x_1 + W_2x_2$, which is tending to bend the beam, as shown in Figure 1.3. This is opposed by the bending resistance of the beam to the left of P.

In the case of the simple cantilever shown in Figure 1.3, the shearing force at P is either $R_1 - W_1$ or $R_2 - W_2$. These terms are equal since they equating upward and downward forces on the beam.

$$R_1 + R_2 = W_1 + W_2$$

The bending moment at P is either $R_1a - W_1x_1$ or $R_2b - W_2x_2$. These terms are equal since they are equating clockwise and counterclockwise moments about P.

$$R_1a + W_2x_2 = R_2b + W_1x_1$$

TABLE 1.1
Standard Bending Cases

Type of Beam	Maximum Moment	Maximum Deflection	Maximum Shear
	$M = PL$ Fixed end	$\delta = \dfrac{PL^3}{3EI}$	$V = P$
	$M = \dfrac{PL}{4}$ Center	$\delta = \dfrac{PL^3}{48EI}$ Center	$V = \dfrac{P}{2}$
	$M = \dfrac{3PL}{16}$	$\delta = \dfrac{PL^3}{48EI\sqrt{5}}$	$V = \dfrac{11}{16}P$
	$M = \dfrac{PL}{2}$ Both ends	$\delta = \dfrac{PL^3}{12EI}$ Guided ends	$V = P$
	$M = \dfrac{PL}{8}$ Center & ends	$\delta = \dfrac{PL^3}{192EI}$ Center	$V = \dfrac{P}{2}$
	$M = \dfrac{PL}{2}$ Fixed ends	$\delta = \dfrac{PL^3}{8EI}$	$V = P$
	$M = \dfrac{PL}{8}$ Center	$\delta = \dfrac{5PL^3}{384EI}$ Center	$V = \dfrac{P}{2}$
	$M = \dfrac{PL}{8}$ Fixed Ends	$\delta = \dfrac{PL^3}{185EI}$	$V = \dfrac{5}{8}P$
	$M = \dfrac{PL}{3}$ Fixed end	$\delta = \dfrac{PL^3}{24EI}$ Guided end	$V = P$
	$M = \dfrac{PL}{12}$ Both ends	$\delta = \dfrac{PL^3}{384EI}$ Center	$V = \dfrac{P}{2}$

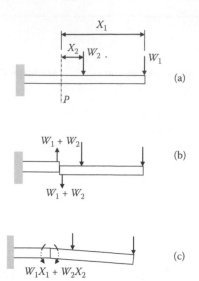

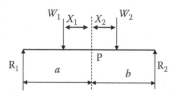

FIGURE 1.3   Shearing forces and bending moments acting on a beam in (a)–(c).

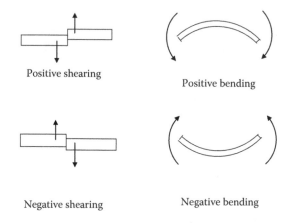

FIGURE 1.4   Simply supported beam.

Positive shearing

Positive bending

Negative shearing

Negative bending

FIGURE 1.5   Conventions used in shearing and bending.

It is immaterial which side of the section is chosen for the calculation of the shearing force or bending moment. When calculating the bending moment due to the distributed load, the part of the load to one side of the section may be considered a concentrated load acting at the center of gravity of that part.

If the bending moment at a section of a beam changes sign, that point is called the point of inflexion or contraflexure.

If the resultant force to the right of a section is upward (or to the left is downward), this is regarded as positive shearing force and the opposite kind of shearing will be regarded as negative (see Figure 1.5).

If the resultant bending moment to the right of the section is clockwise (or to the left is counterclockwise), this will be regarded as a positive bending moment and the opposite kind will be regarded as negative. Thus a positive bending moment causes the beam to bend upward (convex) and the negative bending moment bends it downward (concave).

1.1.4 SHEARING FORCE AND BENDING MOMENT DIAGRAMS

These are diagrams that show the results of the shearing force (SF) and bending moment (BM) along the entire length of the beam. The four most common cases will now be considered.

1.1.4.1 Cantilever with a Concentrated End Load (see Figure 1.6)

$$SF \text{ at } P = -W$$

$$BM \text{ at } P = Wx$$

$$Maximum\ SF = -W$$

$$Maximum\ BM = Wl$$

1.1.4.2 Cantilever with a Uniformly Distributed Load, w, per Unit Length (see Figure 1.7)

$$SF \text{ at } P = -wx$$

$$BM \text{ at } P = wx \cdot \frac{x}{2}$$

$$= \frac{wx^2}{2}$$

$$Maximum\ SF = -wl$$

$$Maximum\ BM = \frac{wl^2}{2}$$

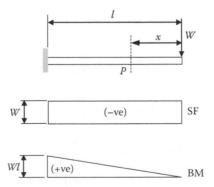

FIGURE 1.6 Cantilever with a concentrated load.

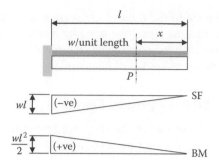

FIGURE 1.7   Cantilever with a uniformly distributed load.

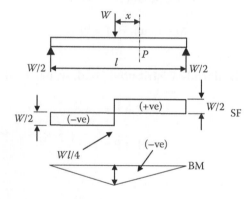

FIGURE 1.8   Simply supported beam with a central load.

1.1.4.3   Simply Supported Beam with Central Concentrated Load (see Figure 1.8)

$$\text{SF at P} = \frac{W}{2}$$

$$\text{BM at P} = \frac{W}{2}\left(\frac{l}{2} - x\right)$$

$$\text{Maximum SF} = \frac{W}{2}$$

$$\text{Maximum BM} = -\frac{W}{2} \cdot \frac{l}{2}$$

$$= -\frac{Wl}{4}$$

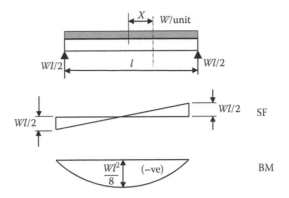

FIGURE 1.9 Simply supported beam with a uniformly distributed load.

1.1.4.4 Simply Supported Beam with Uniformly Distributed Load, w, per Unit Length (see Figure 1.9)

$$\text{SF at P} = \frac{wl}{2} - w\left(\frac{l}{2} - x\right)$$

$$= wx$$

$$\text{BM at P} = \frac{wl}{2}\left(\frac{l}{2} - x\right) + \frac{w}{2}\left(\frac{l}{2} = -x\right)^2$$

$$= \frac{w}{2}\left(\frac{l^2}{4} - x^2\right)$$

$$\text{Maximum SF} = \frac{wl}{2}$$

$$\text{Maximum BM} = -\frac{wl^2}{8}$$

1.2 STRESSES INDUCED BY BENDING

1.2.1 PURE BENDING

If a beam is subjected to a bending moment at each of the beam (see Figure 1.10) the moment is uniform throughout the length of the beam. The fibers in the upper surface of the beam are extended and those in the lower surface are compressed. Tensile and compressive stresses are then induced in the beam, which produces a moment, called the moment of resistance, and is equal and opposite to the applied bending moment.

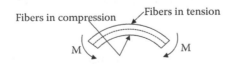

FIGURE 1.10 Surfaces in a section due to a bending moment.

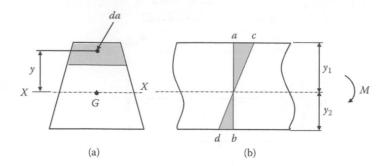

FIGURE 1.11   (a) Represents a cross-section of a beam, to which (b) a bending moment *M* has been applied.

In the theory of bending, which relates the stresses and curvature of the beam to the applied bending moment, the following assumptions are made:

- The beam is initially straight and the radius of curvature is large in comparison to the dimensions of the cross section.
- The material is homogeneous, elastic, and obeys Hooke's law.
- The modulus of material is the same when in tension or compression.
- The stresses are consistent across the depth and width of the section and do not exceed the limits of proportionality.
- A transverse section that is in plane before bending should remain in plane after bending.
- Each longitudinal fiber is unrestrained and free to extend or contract.

1.2.2 STRESS DUE TO BENDING

Figure 1.11 represents a cross section of a beam, to which a bending moment M has been applied; this is acting in a vertical plane through the centroid G. Since the section ab in Figure 1.11(b) remains in plane after bending, it will take up a position cd and the figure acdb will then represent the strain distribution diagram for the section, with ac and bd representing the strains at the top and bottom faces, respectively.

It is evident that the strain will vary linearly from a maximum at the top fibers to a maximum at the bottom fibers, and in doing so changes from tensile to compressive. The plane XX at which the stress becomes zero is termed the neutral plane, and its intersection with a cross section is termed the neutral axis.

Thus the strain at any point, and consequently from Hooke's law, the stress is directly proportional to the distance of that point from the neutral axis.

1.2.3 THE GENERAL BENDING FORMULA

Considering the stresses induced in the beam by the bending moment acting on the beam, and considering further Figure 1.12, let the radius of curvature of the neutral axis at a particular section of the beam be R.

The layer originally of length ce has extended to cd.

$$\therefore \text{strain} = \frac{ed}{cd} = \frac{ed}{ab} = \frac{f_t}{R} \tag{1.2}$$

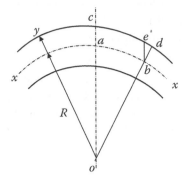

FIGURE 1.12   Stresses induced in a beam due to bending.

using similar triangles:

$$\frac{ed}{bd} = \frac{ab}{ob} = \frac{y}{R}$$ (1.3)

From Equations (1.2) and (1.3)

$$\frac{f_t}{y} = \frac{E}{R}$$ (1.4)

Combining this with Equation (1.2) gives the general bending formula:

$$\frac{M}{I} = \frac{ft}{y} = \frac{E}{R}$$ (1.5)

1.2.4   EXAMPLE

By way of an example consider a beam loaded as in Figure 1.13. Draw a shearing force and bending moment diagram for the figure. If the maximum stress due to bending in the beam is not to exceed 110 MPa, determine the size of a universal beam manufactured to British Standard (BS) 4 rolled steel sections. It is proposed to consider three sizes of sections, 457 × 152 × 74 kg, 406 × 140 × 46 kg, and 356 × 171 × 45 kg (see Table 1.2).

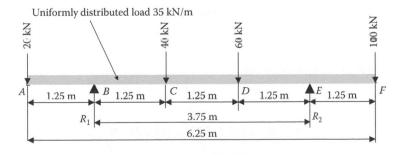

FIGURE 1.13   Uniformly distributed and multiple loads on a beam.

TABLE 1.2
Section Properties

Section	y mm	I m⁴
457 × 152 × 74 kg	228.5	324.53×10^{-6}
406 × 140 × 46 kg	203.0	156.56×10^{-6}
356 × 171 × 45 kg	178.0	120.98×10^{-6}

1.2.4.1 To Determine Reaction Forces

Note that it is good practice to check the results to ensure no mistakes have crept into the calculations due to an oversight (which will always happen despite the best resolve).

Taking moments about R_2

$$3.75 \text{ m} \times R_1 + \left(100 \text{ kN} \times 1.25 \text{ m}\right) + \left(35 \text{ kN} \times 1.25 \text{ m} \times 0.625 \text{ m}\right) = \left(20 \text{ kN} \times 5 \text{ m}\right) + \left(40 \text{ kN} \times 2.5 \text{ m}\right)$$
$$+ \left(60 \text{ kN} \times 1.25 \text{ m}\right)$$
$$+ \left(35 \text{ kN} \times 5 \text{ m} \times 2.5 \text{ m}\right)$$

$$R_1 = \frac{560.1563}{3.75}$$
$$= 149.375 \text{ kN}$$

Taking moments about R_1

$$3.75 \text{ m} \times R_2 + \left(20 \text{ kN} \times 1.25 \text{ m}\right) + \left(35 \text{ kN} \times 1.25 \text{ m} \times 0.625 \text{ m}\right) = \left(40 \text{ kN} \times 1.25 \text{ m}\right) + \left(60 \text{ kN} \times 2.5 \text{ m}\right)$$
$$+ \left(100 \text{ kN} \times 5.0 \text{ m}\right)$$
$$+ \left(35 \text{ kN} \times 5.0 \text{ m} \times 2.5 \text{ m}\right)$$

$$R_2 = \frac{1085.156}{3.75}$$
$$= 289.375 \text{ kN}$$

Check:
Upward forces equal the downward forces.

Upward forces:

$$R_1 + R_2 = 149.375 \text{ kN} + 289.375 \text{ kN}$$
$$= 438.75 \text{ kN}$$

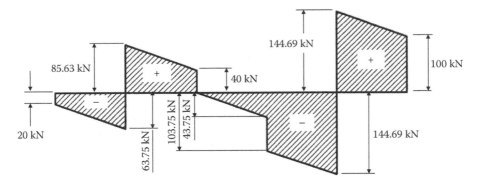

FIGURE 1.14 Shear force diagram.

Downward forces:

$$20 \text{ kN} + 40 \text{ kN} + 60 \text{ kN} + 100 \text{ kN} + \left(35 \text{ kN} \times 6.25 \text{ m}\right)$$

$$= 438.75 \text{ kN}$$

1.2.4.2 Shear Force Diagram

The shear force diagram can now be constructed as shown in Figure 1.14. The maximum shearing force is 144.69 kN at position E. The greatest numerical value will be the maximum bending moment acting on the beam.

As the distributed load covers the entire length of the beam, it is recommended that the bending moment is calculated at 0.625 m intervals (Table 1.3).

TABLE 1.3
Bending Moment Calculations

Section	Distance from A to m	Calculation	Bending Moment
1	0.000		0.00
2	0.625	−(20 kN × 0.625 m) − (35 kN × 0.625 m × 0.625/2)	−19,335.94
3	1.250	−(20 kN × 1.25 m) − (35 kN × 1.25 m × 1.25/2)	−52,343.75
4	1.875	−(20 kN × 1.875 m) − (35 kN × 1.875 m × 1.875/2) + (149.375 kN × 0.625 m)	−5,664.06
5	2.500	−(20 kN × 2.50 m) − (35 kN × 2.50 m × 2.50/2) + (149.375 kN × 1.25 m)	27,343.75
6	3.125	−(20 kN × 3.125 m) − (35 kN × 3.125 m × 3.125/2) + (149.375 kN × 1.875 m) − (40 kN × 0.625 m)	21,679.69
7	3.750	−(20 kN × 3.75 m) − (35 kN × 3.75 m × 3.75/2) + (49.375 kN × 2.50 m) (40 kN × 1.25 m)	2,343.75
8	4.375	−(20 kN × 4.375 m) − (3.5 kN × 4.375 m × 4.375/2) + (149.375 kN × 3.125 m) − (40 kN × 1.875) − (60 kN × 0.625 m)	−30,664.06
9	5.000	−(20 kN × 5.00 m) − (3.5 kN × 5.00 m × 5.00/2) + 149.375 kN × 3.75 m) − (40 kN × 2.50 m) − (6 kN × 1.25 m)	−152,343.75
10	5.625	−(20 kN × 5.625 m) − (3.5 kN × 5.625 m × 5.625/2) + 149.375 kN × 4.375 m) − (40 kN × 3.125 m) − (60 kN × 1.875 m + (289.375 kN × 0.625 m)	−69,335.94
11	6.250	−(20 kN × 6.25 m) − (35 kN × 2.25 m × 6.25/2) + (149.375 kN × 5.00 m) − (40 kN × 3.75 m) − (60 kN × 2.50 m) + (289.375 kN × 1.25 m)	0.00

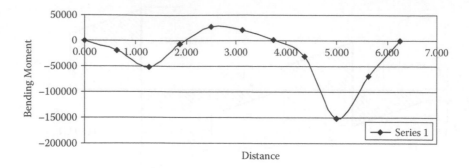

FIGURE 1.15 Bending moment diagram.

TABLE 1.4
Sectional Properties

Section	I m⁴	y mm	Z mm³
457 × 152 × 74 kg	316.47×10^{-6}	228.5	1.385×10^{-3}
406 × 140 × 46 kg	281.16×10^{-6}	203.0	1.385×10^{-3}
356 × 171 × 45 kg	242.38×10^{-6}	178.0	1.361×10^{-3}

From the bending moment diagram (Figure 1.15) it is clearly seen that the maximum bending moment occurs at position E (Figure 1.13) with a value of $-152,343.75$ Nm.

$$M_{max} = f_t \, Z$$

$$\therefore Z = \frac{M}{f_t}$$

$$\therefore Z = \frac{152343.75 \, \text{Nm}}{110 \times 10^6 \, \text{N/m}^2}$$

$$= 1.385 \times 10^{-3} \, \text{m}^3$$

Table 1.4 shows the properties of sections that would match the requirements of the beam problem. Three sections are considered.

From the above calculation it is seen that two sections will be acceptable to meet the strength requirements.

1.2.5 BEST POSITION OF SUPPORTS FOR BEAMS WITH OVERHANGING ENDS

In some cases it is desirable to find the most optimum positions of supports for a beam that overhangs the supports, which gives the least maximum bending stress.

Consider a beam ABCD (Figure 1.16(a) and (b)), which is simply supported at B and C, and both carrying a uniformly distributed load w per unit length throughout. It is possible to adopt a symmetrical arrangement of the supports or to make the overhanging ends of unequal length; the respective bending moment diagram shown in Figure 1.16(b).

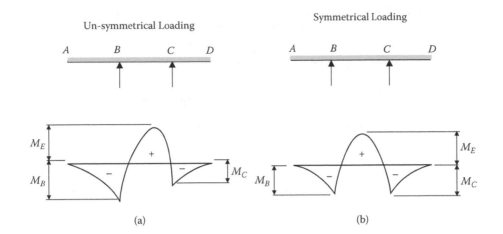

FIGURE 1.16 Beams with overhanging ends.

In either case, let M_E be the maximum positive bending moment and M_B and M_C be the maximum negative bending moments.

When the overhangs are equal,

$$M_B = M_C \qquad\qquad (1.6)$$

If the supports are moved relative to each other a position will be found for which M_B, M_C, and M_E are numerically equal. This arrangement will clearly give the minimum bending moment over the whole length, i.e.,

$$M_E = -M_C = -M_B \qquad\qquad (1.7)$$

For this condition, the overhanging portions will be of equal length and M_E will occur at the middle of the beam.

Let l be the length of the beam and x the overhang at each end (Figure 1.17(a)).

$$M_B = -\frac{1}{2}wx^2$$

$$M_E = R_B\left(\frac{1}{2}l - x\right) - \frac{1}{8}wl^2$$

$$\therefore M_E = \frac{1}{2}wl\left(\frac{1}{2}l - x\right) - \frac{1}{8}wl^2 \qquad\qquad (1.8)$$

$$= \frac{1}{8}wl^2 - \frac{1}{2}wlx$$

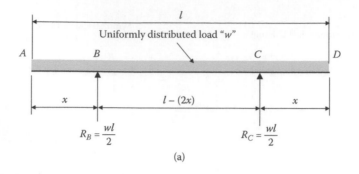

$$R_B = \frac{wl}{2} \qquad\qquad R_C = \frac{wl}{2}$$

(a)

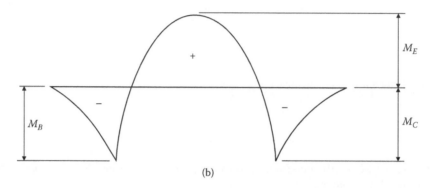

(b)

FIGURE 1.17 Uniformly distributed load acting on a beam with overhanging ends. (a) Beam with a uniformly distributed load with symmetrical overhanging supports. (b) Bending moment diagram for the beam in (a).

$$\text{If } M_E = M_B$$

$$+\frac{1}{2}wx^2 = \frac{1}{8}wl^2 - \frac{1}{2}wlx$$

$$\therefore 4x^2 = l^2 - 4lx$$

$$4x^2 + 4l^2 - l^2 = 0$$

This is a quadratic in x.

$$x = \frac{1}{8}\left[-4\,l \pm \sqrt{\left(16\,l^2 + 16\,l^2\right)}\right]$$

$$= \frac{1}{8}\left(-4\,l \pm \sqrt{32\,l^2}\right)$$

$$= -\frac{1}{2}\,l \pm \frac{1}{2}\,l\sqrt{2}$$

$$= 0.5\,l \pm 0.707\,l$$

Since the negative root is inadmissable

$$x = 0.207l$$

1.3 DEFLECTION IN BEAMS

There are a number of methods to compute the deflection in beams, either simply supported, encastré, cantilever, etc. These include but are not limited to the following:

- Area moment
- Slope deflection
- Moment distribution
- Macaulay's method

1.3.1 AREA MOMENT

The area moment method of analysis is usually attributed to Mohr, who published his method in 1868. It was Professor C.E. Greene of the University of Michigan who in 1872 introduced the principles as they are today. Professor H.F.B. Müller-Brelau extended the method to highly indeterminate structures.

Consider the cantilever shown in Figure 1.18. The cantilever is built in at position A and carries a load at C. Under the action of the load, the cantilever will no longer be horizontal except at position A. The slope of the beam and consequently the deflection will vary from A to C. At position B, for example, a short distance ds from A, the slope will be dθ and the deflection dy.

From the bending equation,

Ignoring the middle term,

$$\frac{M}{I} = \frac{f}{y} = \frac{E}{R} \tag{1.9}$$

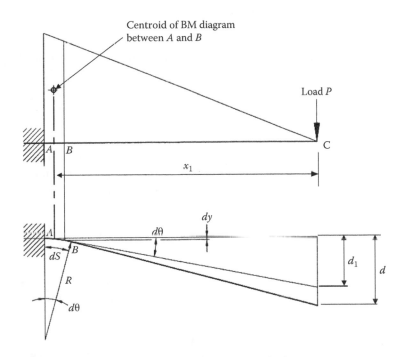

FIGURE 1.18 Analysis of a cantilever using the area-moment method.

$$\frac{M}{EI} = \frac{1}{R} \tag{1.10}$$

where:
   M = the bending moment
   E = the modulus of elasticity
   R = the radius of curvature of the beam

From Figure 1.18 it is seen that

$$dS = R \cdot d\theta \text{ (where } d\theta \text{ is measured in radians).}$$

or

$$d\theta = \frac{dS}{R} = \frac{1}{R} \cdot dS$$

$$= \frac{M \cdot dS}{EI}$$

Consequently the total change in slope from positions A to C will be

$$\theta = \int_{C}^{A} \frac{M \cdot dS}{EI}$$

Returning to the short length dS between A and B, it will be seen that the deflection d_1 at C due to the bending of that short length alone may be found from the following equation:

$$d_1 = d\theta \cdot x_1 = \frac{M \cdot dS}{EI} \tag{1.11}$$

In similar circumstances, where the deflection of a member is in the horizontal direction, the appropriate equation will be

$$d = \int_{C}^{A} \frac{M \cdot y \cdot dS}{EI} \tag{1.12}$$

In Equations (1.11) and (1.12), $M \cdot dS$ is the area of the bending moment diagram in Equation (1.11), x is the lever arm between the centroid of the bending moment diagram and the point of deflection under consideration. From these data two theorems were developed and may be expressed as follows:

Theorem 1

The change in slope between any two points, say A and C in Figure 1.18, in an originally straight beam is equal to the area between corresponding points in the bending moment diagram divided by EI, i.e.,

$$\theta = \frac{\Sigma A}{EI} \tag{1.13}$$

where ΣA is the area of the bending moment diagram.   ∎

Theorem 2

The deflection of a point, say C in Figure 1.18, is an originally straight member under flexure. In the direction perpendicular to the original axis of the member and measured from the tangent at a second point on the member, say A, is equal to the statical moment of the BM diagram divided by EI taken about the first point C, i.e.,

$$d = \frac{\Sigma Ax}{EI} \qquad (1.14)$$

where x is the lever arm between the centroid of the BM diagram and C.   ∎

Consider a beam to demonstrate the application of the above theorems (Figure 1.19), θ_B being the angular change between A and C measured in radians, which is equal to the area of the BM diagram, divided by EI, and d being the deflection, in this case upwards. This is measured perpendicular to the original axis of the member, of the point C from the tangent at A, this being equal to the statical moment of the BM diagram at C, divided by EI.

Now the angle θ_B is equal to the sum of the angles θ_A and θ_C at the ends of the beam. It will be seen that $\theta_A:\theta_C = x:(L - x)$, from which it is seen that the angular change at either end of the beam can be obtained by calculating the support reaction at that end of the beam when it is loaded with its BM diagram divided by EI.

1.3.1.1 Example

Consider a 254 × 146 mm universal beam that is loaded as in Figure 1.20 with a single point load. Take E as 210 GPa and I = 4.427 × 10⁻⁶ m⁴.

Take E as 210 GPa and I = 4.427×10^{-6} m⁴.

Determine the following:

1. The deflection under load
2. The maximum deflection
3. The deflection at the center of the span

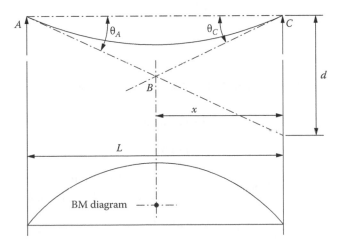

FIGURE 1.19 Beam in Theorem 2.

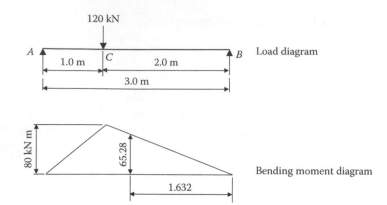

FIGURE 1.20 Load and bending moment diagram for Example 1.3.1.1.

$$\text{BM at C} = \frac{\text{Wab}}{\text{L}}$$

$$= \frac{120 \times 1 \times 2}{3}$$

$$= 80 \text{ kNm}$$

Considering the BM diagram as a load, taking moments about A and rearranging,

$$R_b = \frac{1}{3}\left(\frac{80 \times 2 \times 5}{2 \times 3} + \frac{80 \times 1 \times 2}{2 \times 3}\right)$$

$$= 53.33 \text{ kNm}^2$$

Deflection at the center of the span:

$$\text{BM at center of span} = \frac{1.5}{2} \times 80$$

$$= 60 \text{ kNm}$$

$$\text{Secondary BM} = \left(53.3 \times 1.5\right) - \left(\frac{60 \times 1.5}{2} \times \frac{1.5}{3}\right)$$

$$= 79.95 - 22.5$$

$$= 57.45 \text{ kNm}^3$$

$$\text{Deflection} = \frac{57.45 \times 10^{12} \text{ Nmm}^2}{2.1 \times 10^5 \text{ N/mm}^2 \times 4427 \times 10^4 \text{ mm}^4}$$

$$= 6.18 \text{ mm}$$

Deflection under the load:

$$\text{Secondary BM} = \left(53.3 \times 2\right) - \left(\frac{80 \times 2}{2} \times \frac{2}{3}\right)$$

$$= 106.6 - 53.3$$

$$= 53.33 \text{ kNm}^3$$

$$\text{Deflection} = \frac{53.33 \times 10^{12} \text{ Nmm}^3}{2.1 \times 10^5 \text{ N/mm}^2 \times 4427 \times 10^4 \text{ mm}^4}$$

$$= 5.75 \text{ mm}$$

Maximum deflection:

$$\text{Deflection} = \frac{\text{Secondary Moment}}{EI}$$

It is first necessary to determine the position of the maximum deflection:

The maximum deflection will occur where the secondary M is a maximum. As in all other cases of loading, the maximum BM will occur at a point of zero shear.

The point of zero shear will always occur between the load and center of the span.

Let x = distance of the point x′ of zero secondary shear from B. Then,

$$\text{BM at } X = \frac{x}{2} \cdot 80$$

$$= 40x$$

Therefore,

$$40x \cdot \frac{x}{2} = 20x^2$$

$$= 53.33 \text{ whence } x = 1.632 \text{ m}$$

Secondary BM

$$= \left(53.33 \times 1.632 \times 2\right) - \left(53.33 \times \frac{1.632}{3}\right)$$

$$= \frac{53.33 \times 1.632 \times 2}{3}$$

$$= 57.99 \text{ kNm}^3$$

Maximum deflection:

$$= -\frac{57.99 \times 10^{12}}{2.1 \times 4427 \times 10^9}$$

$$= 6.238 \text{ mm}$$

The slopes at the beam ends will now be considered.

Slope at end B:

$$\text{Slope at X} = 0$$

$$\text{BM at X} = \frac{80 \times 1.632}{2}$$

$$= 65.28 \text{ kNm}$$

$$\text{Area of BM diagram between X and B} = \frac{65.28 \times 1.632}{2}$$

$$= 53.33 \text{ kNm}^2$$

$$\text{Slope at B} = \frac{53.33}{EI}$$

$$= +\frac{53.33 \times 10^9}{2.1 \times 10^5 \times 4427 \times 104}$$

$$= 0.00574 \text{ radians}$$

Slope at end A:

$$\text{The total area of the BM diagram} = \frac{80 \times 3}{2}$$

$$= 120 \text{ kNm}^2$$

Consequently, the area of the BM diagram between X and A:

$$= 120 - 53.33$$

$$= 66.67 \text{ kNm}^2$$

Therefore,

$$\text{Slope at A} = -\frac{66.7}{EI}$$

$$= -\frac{66.7 \times 10^9}{2.1 \times 10^5 \times 4427 \times 10^4}$$

$$= -0.00717 \text{ radian}$$

The result is negative, as the area is measured to the left of X.

Slope at C:

The area of the BM diagram between X and C = $\dfrac{80 \times 65.28}{2} \times (2 - 1.632)$

$$= 26.7 \text{ kNm}^2$$

C is to the left of X; therefore:

$$\text{Slope at C} = -\frac{26.7 \times 10^9}{EI}$$

$$= \frac{26.7 \times 10^9}{2.1 \times 10^5 \times 4427 \times 10^4}$$

$$= 0.00287 \text{ radian}$$

1.3.2 SLOPE DEFLECTION

In the method of slope deflection, joint rotations and deflections are treated as unknown quantities, and once they have been evaluated, the moments follow automatically by substituting these values in standards equations.

Consider the member AB in Figure 1.21, in which the member is one unloaded span of a continuous beam and is of a uniform moment of inertia.

For the conditions shown in Figure 1.21,

$$M_{AB} = 2EK(2\theta_A + \theta_B - 3R)$$

$$M_{BA} = 2EK(2\theta_B + \theta_A - 3R)$$

where
E = modulus of elasticity (210×10^9 N/m^2 for steel)
K = stiffness factor (I/L) (for the particular beam in question)
θ_A and θ_B are angles the joints make to the horizontal
R is the angle of rotation of B with respect to A when B sinks the amount of d (i.e., d/L)

With regard to sign conventions,
θ is positive when the tangent to the beam rotates in a clockwise direction.
R is positive when the beam rotates in a clockwise direction.
M is positive when the moment acts in a clockwise direction.

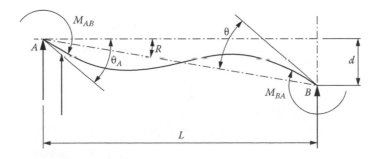

FIGURE 1.21 Slope–deflection symbols.

$$M_{BA} = 2EK\left(2\theta_A + \theta_B - 3R\right) - FEM_{AB}$$

$$M_{BA} = 2EK\left(2\theta_B + \theta_A - 3R\right) + FEM_{BA}$$

Consider that the span AB carries a load acting downward in the normal fashion: FEM_{AB} and FEM_{BA} are fixed end moments that would exist if the member AB were a fixed end beam.

A selection of fixed-end moments (FEM) is shown in Table 1.3.

When end A of a beam AB is hinged, the moments for the other end are modified as follows:

$$M_{BA} = 2EK\left(3\theta_B - 3R\right)\left(\text{unloaded condition}\right)$$

$$M_{BA} = 2EK\left(3\theta_B + \theta_A - 3R\right) + FEM_{BA}\left(\text{loaded condition}\right)$$

When calculating the values of FEM's for loads acting downward in the normal fashion the appropriate signs can be ignored, as the fundamental formulas provide the correct signs automatically.

The final BM diagram is prepared by considering all hogging moments as negative and all sagging moments as positive.

When the slope deflection method is used to find the moments in a continuous beam, the slope of the beam over each internal support is calculated. The values of the slopes may be useful in calculating the deflections in interior spans, but care needs to be exercised with signs. In the slope deflection calculations a positive value for the slope means the beam has rotated in a clockwise direction.

Furthermore, it is essential the values used in the calculations should be consistent with the values used for E and I; otherwise, the values for the slopes will not relate to the units used in the deflection calculations.

By way of an example, see Example 1.1.

Example 1.1

Consider the continuous beam in Figure 1.22 under the action of the depicted loads; the joint B rotates in a counterclockwise direction. With the calculation of the rotation θ_B the whole beam can be resolved.

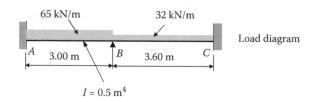

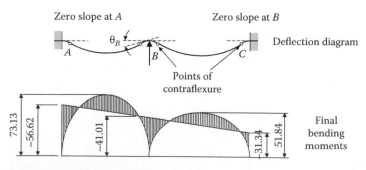

FIGURE 1.22   Example of a continuous beam.

Let the suffixes 1 and 2 be assigned to AB and BC, respectively.

A and C are fixed in direction as well as position.

Therefore:

$$\theta_A = 0 = \theta_C$$

Points A, B, and C are on the same level, therefore

$$FEM_{AB} = FEM_{BA}$$

$$= \frac{wL}{12}$$

$$= 48.75 \, kNm$$

$$FEM_{BC} = FEM_{CB}$$

$$= \frac{wL}{12}$$

$$= 34.56 \, kNm$$

Now

$$M_{BA} + M_{BC} = 0$$

Hence

$$2EK_1\left(2\theta_B + \theta_A - 3R_1\right) + 48.75 + 2EK_2\left(2\theta_B + \theta_C - 3R_2\right) - 34.56 = 0$$

But

$$\theta_A - 0 = \theta_c$$

$$R_1 = 0 = R_2$$

$$K_1 = \frac{I}{L} = \frac{5000 \times 10^{-8}}{3.0} \frac{m^4}{m}$$

and

$$K_2 = \frac{I}{L} = \frac{5000 \times 10^{-8}}{3.6} \frac{m^4}{m}$$

Therefore

$$\left(2E \times \frac{5000 \times 10^{-8}}{3} \times 2\theta_B\right) + \left(2E \times \frac{5000 \times 10^{-8}}{3.6} \times 2\theta_B\right) = 34.56 - 48.75$$

$$6.667 \times 10^{-5} \, E\theta_B + 5.556 \times 10^{-5} \, E\theta_B = 34.56 - 48.75$$

Now

$$E\theta_b\left(6.667 \times 10^{-5} + 5.556 \times 10^{-5}\right) = -14.19$$

$$E\theta_B = \frac{-14.19}{0.1222 \times 10^3}$$

$$E\theta_B = -116.121 \times 10^3$$

Using the basic formula

$$M_{BA} = M_{BC}$$

$$= 2EK_1\left(2\theta_B\right) + 48.75$$

$$= \left(-2 \times \frac{5000 \times 10^{-8}}{3.0} \times 2 \times 116.121 \times 10^3\right) + 48.75$$

$$= 41.0086 \text{ kNm}$$

$$M_{CB} = 2EK_2\left(\theta_B\right) + 34.56$$

$$= \left(-2 \times \frac{5000 \times 10^{-8}}{3.6} \times 116.121 \times 10^3\right) + 34.56$$

$$= 31.3344 \text{ kNm}$$

Example 1.2

In this example, a continuous beam ABCDE involves the treatment of simply supported end spans and a cantilever. (see Figure 1.23).

Let the suffixes 1, 2, and 3 apply to AB, BC, and CD, respectively.

The supports A, B, C, and D are all on the same level.

Hence, $\qquad\qquad\qquad\qquad R_1 = 0 = R_2 = R_3$

Also, $\qquad\qquad\qquad\qquad M_{BA} + M_{BC} = 0$

Hence, $\qquad\qquad\qquad\qquad M_{CB} + M_{CD} = 0$

$$M_{DC} = 60 \text{ kNm}$$

The effect at C due to the cantilever DE is that FEM_{CD} is reduced in value by half the amount of M_{DC}, i.e., 30 kNm.

Now:

$$EK_1\left(3\theta_B - 3R_1\right) + FEM_{BA} + 2EK_2\left(2\theta_B + \theta_C - 3R_2\right) - FEM_{BC} = 0$$

$$2EK_2\left(2\theta_C + \theta_B - 3R_2\right) + FEM_{CB} + EK_3\left(3\theta_C - 3R_3\right) - FEM_{CD} + M_{DC}/2 = 0$$

$$E \times 1.89 \times 10^{-5}\left(3\theta_B\right) + 101.25 + 2E \times 1.733 \times 10^{-5}\left(2\theta_B + \theta_C\right) - 58.33 = 0$$

$$12.6 \times 10^{-5} E\theta_B + 3.466 \times 10^{-5} E\theta_C + 42.92 = 0 \quad \ldots\text{(i)}$$

$$2E \times 1.733 \times 10^{-5}\left(2\theta_C + \theta_B\right) + 58.33 + E \times 1.467 \times 10^{-5}\left(3\theta_C\right) - 103.1 + 30 = 0$$

$$3.466 \times E\theta_B \times 10^{-5} + 11.333E\theta_C \times 10^{-5} - 14.77 = 0$$

or $\qquad\qquad\qquad\qquad 12.6E\theta_B \times 10^{-5} + 41.199E\theta_C \times 10^{-5} - 53.69 = 0 \quad \ldots\text{(ii)}$

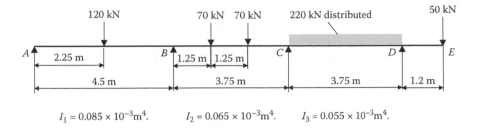

120 kN   70 kN   70 kN   220 kN distributed   50 kN

A   2.25 m   B   1.25 m | 1.25 m   C   D   E

4.5 m   3.75 m   3.75 m   1.2 m

$I_1 = 0.085 \times 10^{-3} m^4.$ $I_2 = 0.065 \times 10^{-3} m^4.$ $I_3 = 0.055 \times 10^{-3} m^4.$

Deflection diagram

Final bending moments

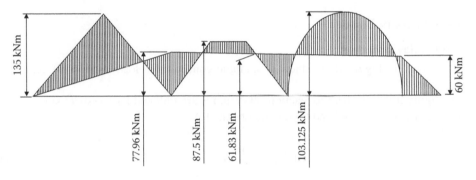

FIGURE 1.23 Loading, deflection, and bending moment diagram for Example 1.2.

Subtracting (i) from (ii)

$$37.733E\theta_C \times 10^{-5} = 96.61$$
$$E\theta_C = 2.56 \times 10^5$$

Substituting in Equation (i),

$$E\theta_B = -4.11 \times 10^5$$

Employing these values of $E\theta_B$ and $E\theta_C$,

$$M_{BA} = -M_{BC}$$
$$= EK_1\left(3\theta_B\right) + 101.25$$
$$= -1.89 \times 10^{-5} \times 3 \times 4.11 \times 10^5 + 101.25$$
$$= 77.96 \, kNm$$

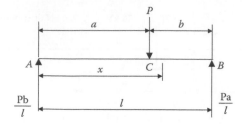

FIGURE 1.24 Simply supported beam with a single offset concentrated load.

$$M_{CB} = -M_{CD}$$

$$= 2EK_2\left(2\theta_C + \theta_B\right) + 58.33$$

$$= 2 \times 1.733 \times 10^{-5}\left(2.56 \times 2 - 4.11\right) \times 10^{-5} + 58.33$$

$$= 61.83\,\text{kNm}$$

1.3.3 DEFLECTION IN BEAMS

1.3.3.1 Macaulay's Method

Consider the beam in Figure 1.24 subject to a single concentrated load applied off center to the beam.

From the figure it is seen that the reactions at A and B are Pb/l and Pa/l, respectively.

Considering the first part AC and taking the origin at A,

$$EI\frac{d^2y}{dx^2} = -\frac{Pb}{l}x \tag{1.15}$$

$$EI\frac{dy}{dx} = -\frac{Pb}{l}\frac{x^2}{2} + A_1 \tag{1.16}$$

and

$$EI = -\frac{Pb}{l}\frac{x^3}{6} + A_1 x + B_1 \tag{1.17}$$

When $x = 0$, $y = 0$, so that $B_1 = 0$.

The point at which the slope is zero is unknown and the condition that $y = 0$ when $x = 0$ cannot be used as the equation do not apply beyond point C. Therefore, the constant of integration A_1 must remain unknown for the present.

Now consider the part CB still keeping the origin at point A.

$$EI\frac{d^2y}{dx^2} = -\frac{Pb}{l}x + P[x - a] \tag{1.18}$$

$$EI\frac{dy}{dx} = -\frac{Pb}{l}\frac{x^2}{2} + P\left[\frac{x^2}{2} - ax\right] + A_2 \tag{1.19}$$

and

$$EI = -\frac{PB}{l}\frac{x^3}{6} + P\left[\frac{x^3}{6} - a\frac{x^2}{2}\right] + A_1x + B_1$$

(1.20)

When $x = 0$, $y = 0$, so that $B_2 = \frac{Pal^2}{3} - A_2l$

For reasons similar to those given above, the constant of integration cannot be determined at this stage.

Equating the slope at C is given by Equations (1.16) and (1.19) when $x = a$:

$$-\frac{Pb}{l}\frac{a^2}{2} + A_1 = -\frac{Pb}{l}\frac{a^2}{2} + P\left[\frac{a^2}{2} - a^2\right] + A_2$$

Equating the deflection at C is given by Equations (1.17) and (1.19) when $x = a$,

$$-\frac{Pb}{l}\frac{a^3}{6} + A_1a = -\frac{Pb}{l}\frac{a^3}{6} + P\left[\frac{a^3}{6} - \frac{a^3}{2}\right] + A_2a + \left(\frac{Pal^2}{3} - A_2l\right)$$

A_1 and A_2 may be determined from these two equations, and hence the slope and deflection can be obtained at any point using Equations (1.16) and (1.17) for $x < a$ and Equations (1.18) and (1.19) for $x > a$.

For two concentrated loads, three sets of equations will be required for the three ranges of beam. This leads to six constants of integration. Two of these are obtained from the conditions that $y = 0$ at $x = 0$ and $y = 1$, and the remaining four are resolved by equating slopes and deflections under the loads.

It is clearly seen that this method is extremely cumbersome, and it should be noted that the constants of integration will be different for each section of the beam.

Referring to Equation (1.18), this could legitimately be integrated as follows:

$$EI\frac{dy}{dx} = -\frac{Pb}{l}\frac{x^2}{2} + \frac{P}{2}[x-a]^2 + A_2'$$

(1.21)

$$EIy = -\frac{Pb}{l}\frac{x^3}{6} + \frac{P}{6}[x-a]^3 + A_2'x + B_2'$$

(1.22)

The constant A_2 will not be the same as A_2' previously obtained $\left(A_2 = A_2' + \frac{a^2}{2}\right)$

and B_2' will not be the same as $B_2 \left(b_2 = B_2' - \frac{a^3}{6}\right)$

If the slopes at C are now equated using Equations 1.16 and 1.19 when $x = a$

$$-\frac{Pb}{l}\frac{a^2}{2} + A_1 = -\frac{Pb}{l}\frac{a^2}{2} + \frac{P}{2}[a-a]^2 + A_2'$$

$$\therefore A_1 = A_2' (= A)$$

Similarly, equating deflections at C, using Equations (1.17) and (1.22) when x = a,

$$-\frac{Pb}{l}\frac{a^3}{6}+Aa+B_1 = -\frac{Pb}{l}\frac{a^3}{6}+\frac{P}{6}[a-a]^3+Aa+B_2'$$

$$\therefore B_1 = B_2'(=B)$$

Thus, by this method of integration, the constants of integration for each section of the beam are the same. There is a further advantage in that Equations (1.18), (1.21), and (1.22) are identical with Equations (1.15), (1.16), and (1.17), except for the additional term involving [x – a], which only comes in when x > a, i.e., when [x – a] is positive.

Thus Equations (1.18), (1.21), and (1.22) may be regarded as applying to the complete beam provided that, for any value of x that makes [x – a] negative, this term is ignored.

This method is known as *Macaulay's method*, and it is convenient to use square brackets for terms such as [x – a], which have to be treated in this special manner.

Now proceeding with this case, the deflection Equation (1.22) can be simplified to:

$$EIy = -\frac{Pb}{l}\frac{x^3}{6}+\frac{P}{6}[x-a]^3+Ax+B$$

When x = l, y = 0, so that $A = \frac{Pab}{6l}(l+b)$

this can therefore be ignored.

When x = 0, y = 0, so that B = 0 since [x – a]^3 is negative for this value of x and the term involving this can therefore be ignored.

The deflection under load is given by

$$EIy = -\frac{Pb}{l}\frac{x^3}{6}+\frac{Pb}{6l}(l+b)a$$

$$\therefore y = \frac{Pa^2b^2\ *}{3EIl} \tag{1.23}$$

$$*\text{When } a = b = \frac{l}{2} \text{ this will reduce down to } \frac{PL^3}{48EI}$$

The maximum deflection will occur between the load points and the center of the beam. If a > b, this point will correspond to x < a, so that from Equation (1.21)

$$-\frac{Pb}{l}\frac{x^2}{2}+\frac{Wab}{6l}(l+b)=0$$

Writing a = l – b, this will reduce to

$$x = \sqrt{\left(\frac{l^2-b^2}{3}\right)}$$

Now substituting in Equation (1.22)

$$EIy_{max} = -\frac{Pb}{6l}\left(\frac{l^2-b^2}{3}\right)^{3/2} + \frac{Pab}{6l}(l+b)\sqrt{\left(\frac{l^2-b^2}{3}\right)}$$

or (1.24)

$$y_{max} = \frac{Pb\left(l^2-b^2\right)}{9\sqrt{3}EIl}$$

As $b \to 0$, $x \to l/\sqrt{3}$, which is approximately $l/13$ from the center of the beam.

Thus the maximum deflection is very close to the center of the beam, even for an extremely unsymmetrical load, and for the most normal cases of loads on beams simply supported at both ends, the maximum deflection will be virtually identical with the central deflection.

1.4 SHEAR DEFLECTION IN BEAMS

1.4.1 Introduction

Shear deflection in beams is usually negligible in metal beams unless the span-to-depth ratio is extremely small, i.e., where shear stresses tend to be high. For beams where $l/d \geq 10.0$ shear deflection may be ignored. Where $l/d \geq 3$ use the following procedure.

Deflection due to shear forces on the beam section is given by the following equation:

$$\delta_s = F \cdot \int \frac{V \cdot v}{A \cdot G} \cdot dx \tag{1.25}$$

where:

V = vertical shear at section due to actual load
v = vertical shear at section due to unit load at section
A = area of section
G = shear modulus
F = form factor for shear deflection (see Table 1.5)
y_1 = distance from neutral axis to nearest surface of flange
y_2 = distance from neutral axis to extreme fiber
t_1 = thickness of web (or webs in box beams)
t_2 = width of flange
k = radius of gyration

TABLE 1.5
Form Factor for Shear Deflection

Shape (Solid Sections)	Form factor (F)
Rectangular, triangular, trapezoid	5/6
Diamond	31/30
Circular	10/9
Thin tube (circular)	2
I or box section (flanges and webs of uniform thickness)	$\left[1+\dfrac{3\left(y_2^2-y_1^2\right)y_1}{2y_2^3}\dfrac{t_1}{t_1}-1\right]\dfrac{4y_2^2}{10k^2}$
I beam, A = area of web (approx.)	1

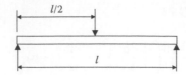

FIGURE 1.25   Simply supported beam with a central load.

1.4.2   DETERMINE THE SHEAR DEFLECTION IN A SIMPLY SUPPORTED BEAM WITH A CENTRAL POINT LOAD

The shear deflection (δ_s) may be found using one of the following relationships:

$$\delta_s = F \int \frac{V_v}{AG} \cdot d_x \qquad (1.26)$$

$$\delta_s = \frac{F}{AG} \cdot \frac{P}{2} \cdot \frac{l}{2} \cdot \int dx \qquad (1.27)$$

$$\delta_s = \frac{1}{4} \cdot F \cdot \frac{Pl}{AG} \qquad (1.28)$$

1.4.3   SHEAR DEFLECTION OF SHORT BEAMS

In cases where $l/d < 3.0$, the assumption of linear stress distribution on which the simple theory of flexure is based is no longer valid for short beams. The equation given in (1.25) will give a relatively accurate result for $l/d \geq 3.0$.

With $l/d < 3.0$ the stress distribution will change radically, and dependent upon the loading and support, the maximum stress may be greater than the engineering prediction for the theory of bending stresses (My/I).

In the section that follows, the ratio of the actual stress to the engineering theory is plotted against span/depth for a uniformly distributed load.

The distribution is considered over:

- The entire beam (Figures 1.26 and 1.27)
- The center portion (Figures 1.28 and 1.29)

1.4.4   SHORT BEAM WITH UNIFORMLY DISTRIBUTED LOAD OVER THE ENTIRE SPAN

Considering the ratio of the maximum stress to My/I stresses,

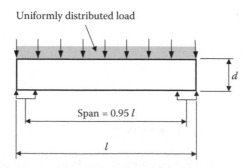

FIGURE 1.26   Short beam with a distributed load over the entire span.

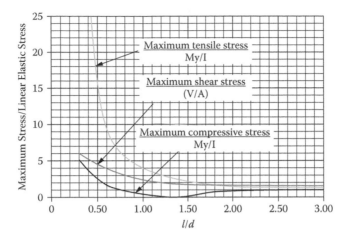

FIGURE 1.27   Graph for Figure 1.26 with *l*/d < 3.0. Uniform load over entire length.

1.4.5   SHEAR DEFLECTION IN A SHORT BEAM WITH UNIFORMLY DISTRIBUTED LOAD OVER CENTER SPAN

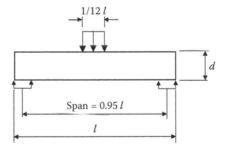

FIGURE 1.28   Uniform load over central portion of beam.

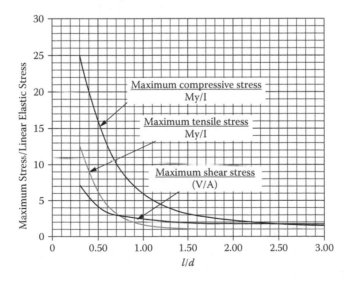

FIGURE 1.29   Graph of Figure 1.28 with *l*/d < 3.0.

1.5   PROPERTIES OF A PLANE AREA

1.5.1   NOTATION

δA = Incremental area

A = Area

o = Datum, point of origin

CG = Centroid, center of gravity

x, y = Distance along x and y axis from origin

$\bar{x}, \bar{y}$ = Distance of centroid from origin

x_o, y_o = Distance from centroid to shear center

A_x, A_y = First moment of area

I_x, I_y = Second moment of area

I_{xy} = Product of inertia

I_p = Polar moment of inertia

k = Radius of gyration

I_x', I_y' = Principle moments of inertia

θ = Axis of princial axis to datum axis

e = Distance to shear center

J = Torsional constant

Γ = Warping constant

1.5.2   GENERAL DEFINITIONS (SEE FIGURE 1.30)

1.5.2.1   Area

The area may be defined as the quantity of a two-dimensional space a body occupies.

$$A = \int \cdot \delta A \tag{1.29}$$

1.5.2.2   First Moment of Area

The first moment of area of a section is a measure of the distribution of the area on a section. It is usually used to find the position of the centroid of the section.

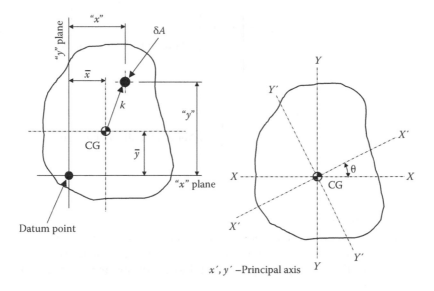

FIGURE 1.30 Moment of inertia–general definitions.

$$Ay = \int \delta A \cdot y \text{ about the x axis.}$$

$$Ax = \int \delta A \cdot x \text{ about the x axis.}$$

$$Ax = Ay = 0 \text{ about the centroid.}$$

1.5.2.3 Centroid: Center of Area

The centroid may be defined as that point in the plane of the area through which the moment of the area is zero.

The axis through which the section bends also passes through this point.

The position of the centroid is

$$\bar{y} = \frac{A \cdot y}{A} \qquad \bar{x} = \frac{A \cdot x}{A} \tag{1.30}$$

where $A \cdot x$ and $A \cdot y$ = the first moment of area.

Note that the center of area coincides with the center of gravity of the area represented as an infinitely thin homogeneous plate.

1.5.2.4 Second Moment of Area

The second moments of area, or moments of inertia of a section, are quantities used in defining the section's ability in resisting bending actions.

The values calculated about an arbitrary axis are usually used only as a means of finding the values about a centroid.

$$I_{xx} = \int \delta A \cdot y^2 \left.\right\rbrace \text{About axis through origin} \qquad (1.31)$$

$$I_{yy} = \int \delta A \cdot x^2 \qquad (1.32)$$

$$I_x = I_{xx} - A\bar{y}^2 \left.\right\rbrace \text{About axis through origin} \qquad (1.33)$$

$$P_y = I_{yy} - A\bar{x}^2 \qquad (1.34)$$

1.5.2.5 Product of Inertia

The product of inertia is with respect to a pair of rectangular axes in its plane. The sum of the products obtained by multiplying each element of the area δA by its coordinates with respect to those axes x and y is therefore the quantity $\int \delta A xy$. It is useful when transposing other terms to a different axis.

$$I_{xy} = \int \delta A \cdot x \cdot y \quad \text{About axis through origin} \qquad (1.35)$$

1.5.2.6 Polar Moment of Inertia

The polar moment of inertia can be defined as the moment of inertia of an area with respect to a point on its surface. It is equal to the sum of the moments of inertia with respect to any axes in the plane of the area, at right angles to each other and passing through the point of intersection of the polar axis with the plane. This term is a measure of the section's ability to resist torque.

$$I_p = \int \delta A \cdot r^2 \qquad (1.36)$$

$$I_p = \int \delta A \cdot \left(x^2 + y^2\right) \qquad (1.37)$$

$$I_p = I_x + I_y \qquad (1.38)$$

1.5.2.7 Radius of Gyration

The radius of gyration of a section is the distance from the inertia axis to that point where, if the entire mass could be concentrated at that point, its moment of inertia would remain the same.

$$Ak^2 = I$$

$$k = \left[\frac{I_x}{A}\right]^{\frac{1}{2}} \qquad (1.39)$$

$$kx = \left[\frac{I_x}{A}\right]^{\frac{1}{2}}, \qquad ky = \left[\frac{I_y}{A}\right]^{\frac{1}{2}}, \qquad kz = \left[\frac{I_p}{A}\right]^{\frac{1}{2}}$$

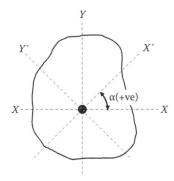

FIGURE 1.31   Moments of inertia about an inclined plane.

1.5.2.8   Moment of Inertia about Inclined Axes (see Figure 1.31)

$$I_{x'} = I_{xx} \cos^2 \alpha - I_{xy} \sin 2\alpha + I_{yy} \sin^2 \alpha \qquad (1.40)$$

$$I_{y'} = I_{xx} \sin^2 \alpha - I_{xy} \sin 2\alpha + I_{yy} \cos^2 \alpha \qquad (1.41)$$

$$I_{x'y'} = I_{xy} \cos 2\alpha + \left(\frac{I_{xx} - I_{yy}}{2} \right) \sin 2\alpha \qquad (1.42)$$

1.5.2.9   Principal Axes

In problems involving unsymmetrical bending, the moment of an area is frequently used with respect to a certain axis called the principal axis. A principal axis of an area is an axis about which the moment of inertia of the area is either greater than or less than that for any other axis passing through the centroid of the area. Axes about which the product of inertia is zero are the principal axes.

Since the product of inertia is zero about symmetrical axes, it follows that the symmetrical axes are the principal axes (for symmetrical sections). The angle between a set of rectangular centroidal axes and the principal axes is given by

$$\theta = \frac{1}{2} \arctan\left[\frac{2 \cdot I_{xy}}{I_x - I_y} \right] \qquad (1.43)$$

1.5.2.10   Principal Moments of Inertia

The principal moments of inertia are second moments of area about the principal axes. The relationship between these and the moments of inertia through the centroid are as follows:

$$I_{x'} = \frac{1}{2}\left(I_x + I_y\right) + \left[\frac{1}{4}\left(I_y - I_x\right)^2 + I_{xy}^2 \right]^{\frac{1}{2}}$$

$$I_{y'} = \frac{1}{2}\left(I_x + I_y\right) - \left[\frac{1}{4}\left(I_y - I_x\right)^2 + I_{xy}^2 \right]^{\frac{1}{2}}$$

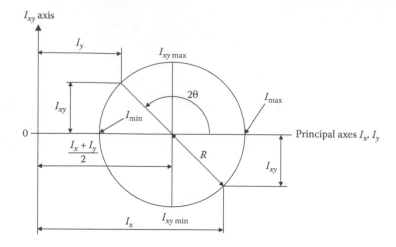

FIGURE 1.32   Features of the Mohr's circle.

1.5.2.11   Mohr's Circle for Moment of Inertia (see Figure 1.32)

A Mohr circle may be constructed to give the location of the principal axes and the values of the principal moments of inertia.

Given I_x, I_y, and I_{xy} about the center of gravity,

$$R = \left[\left[\frac{I_x - I_y}{2}\right]^2 + I_{xy}^2\right]^{\frac{1}{2}}$$

I_{xy} about a principal axes is zero.

θ angle between the principal and the x axis.   (1.44)

I_{max}, I_{min} moments of inertia about the prinipal axis.

The following statements and equations are applicable:

- Any axis of symmetry is one of the principal axes.
- The products of inertia are zero if one of the axes is an axis of symmetry.
- The product of inertia through the centroid is zero.
- The moments of inertia and radius of gyration w.r.t. the neutral axis are less than those for any other parallel axis.
- $I_p = I_{ox} + I_{oy}$ (1.45)
- $I_p = I_x + I_y$
- $I_p = I_{x'} + I_{y'}$
- $I_\theta = I_{ox} \cos^2 \theta + I_{oy} \sin^2 \theta - I_{xy} \sin 2\theta$

1.5.3   Torsional Constant (J)

The torsional constant, J, of a member of uniform cross section is the geometric constant by which the shear modulus, G, must be multiplied to obtain the torsional rigidity of the member. The torsional rigidity is defined as the factor by which a torque, T, applied to each end of the member has to be divided in order to obtain the twist per unit length. It is assumed that the ends of the member are not axially constrained and can be expressed by the equation

$$\varphi = \text{twist per unit length} = \frac{T}{G.\,J}$$ (1.46)

Thus the torsional constant J is dependent on the form and dimensions of the cross section. For circular sections (St. Venant), $J = I_p$ (the polar moment of inertia); for other sections it may be much less than I_p.

1.5.3.1   For Solid Sections

$$J = \frac{A^4}{4\pi^2 I_p}$$

Area = Area of section   (1.47)

FIGURE 1.33   Torsional constant for a solid section. I_p = Polar moment of section

1.5.3.2   For Closed Sections

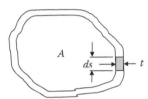

$$J = \frac{4A^2}{\int \dfrac{ds}{t}}$$

s = peripheral length

t = thickness   (1.48)

FIGURE 1.34   Torsional constant for a closed section.   A = enclosed area

1.5.3.3   For Open Sections

The value of the torsional constant J for open sections, which can be idealized to individual rectangular elements, is given by the following formulae:

$$J = \sum c_i \cdot b_i \cdot t_i^3 \qquad (1.49)$$

where

$$c_i = \left[\frac{1}{3} - \frac{0.21 \cdot t}{b} \left(1 - \frac{t^4}{12 \cdot b^4} \right) \right]$$

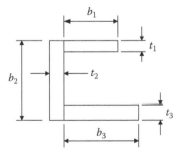

FIGURE 1.35   Torsional constant for an open section.

1.5.4   SECTION PROPERTY TABLES (SEE TABLE 1.6)

TABLE 1.6
Section Property Tables for Various Sections

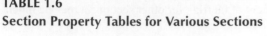

Square	Hollow Square

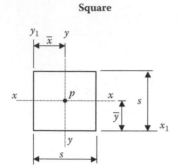
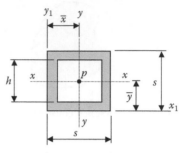

	Square	Hollow Square
Area	s^2	$s^2 - h^2$
Centroid	$\bar{x} = \bar{y} = \dfrac{s}{2}$	$\bar{x} = \bar{y} = \dfrac{s}{2}$

Moment of Inertia

Square:
$$I_x = I_y = \frac{s^4}{12}$$
$$I_{x1} = I_{y1} = \frac{s^4}{3}$$
$$I_p = I_x + I_y = \frac{s^4}{6}$$
$$I_{p1} = I_{x1} + I_{y1} = \frac{2s^4}{3}$$

Hollow Square:
$$I_x = I_y = \frac{s^4 - h^4}{12}$$
$$I_{x1} = I_{y1} = \frac{4s^4 - 3s^2 h^2 - h^4}{12}$$
$$I_p = I_x + I_y = \frac{s^4 - h4}{6}$$
$$I_{p1} = I_{x1} + I_{y1} = \frac{4s^4 - 3s^2 h^2 - h^4}{6}$$

Radius of Gyration

Square:
$$k_x = k_y = 0.289s$$
$$k_{x1} = k_{y1} = 0.577s$$
$$k_p = 0.408s$$
$$k_{p1} = 0.816s$$

Hollow Square:
$$k_x = k_y = 0.289\sqrt{s^2 + h^2}$$
$$k_{x1} = k_{y1} = 0.289\sqrt{4s^2 + h^2}$$
$$k_p = 0.408\sqrt{s^2 + h^2}$$
$$k_{p1} = 0.408\sqrt{4s^2 + h^2}$$

TABLE 1.6 *(Continued)*
Section Property Tables for Various Sections

Rectangle	Hollow Rectangle
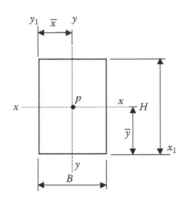	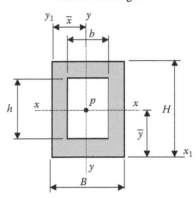

Area

$$BH \qquad BH - bh$$

Centroid

$$\bar{x} = \frac{1}{2}B \qquad \bar{y} = \frac{1}{2}H \qquad\qquad \bar{x} = \frac{1}{2}B \qquad \bar{y} = \frac{1}{2}H$$

Moment of Inertia

$$I_x = \frac{BH^3}{12} \qquad I_y = \frac{HB^3}{12}$$

$$I_{x1} = \frac{BH^3}{3} \qquad I_{y1} = \frac{HB^3}{3}$$

$$I_p = \frac{BH}{12}\left(H^2 + B^2\right)$$

$$I_{p1} = \frac{BH}{3}\left(H^2 + B^2\right)$$

$$I_x = \frac{\left(BH^3 - bh^3\right)}{12} \qquad I_y = \frac{\left(HB^3 - hb^3\right)}{12}$$

$$I_{x1} = \frac{BH^3}{3} - \frac{bh\left(3H^2 + h^2\right)}{12}$$

$$I_{y1} = \frac{1}{3}HB^3 - \frac{bh\left(3B^2 + b^2\right)}{12}$$

$$I_p = I_x + I_y \qquad I_{p1} = I_{x1} + I_{y1}$$

Radius of Gyration

$$k_x = 0.289H \qquad K_y = 0.289B$$

$$k_{x1} = 0.577H \qquad K_{y1} = 0.577B$$

$$k_p = 0.289\sqrt{H^2 + B^2}$$

$$k_{p1} = 0.577\sqrt{H^2 + B^2}$$

$$k_x = \sqrt{\frac{BH^3 - bh^3}{12\left(BH - bh\right)}} \qquad k_{x1} = \sqrt{\frac{I_{x1}}{BH - bh}}$$

$$k_y = \sqrt{\frac{HB^3 - hb^3}{12\left(BH - bh\right)}} \qquad k_{y1} = \sqrt{\frac{I_{y1}}{BH - bh}}$$

$$k_p = \sqrt{\frac{I_p}{BH - bh}} \qquad k_{p1} = \sqrt{\frac{I_{p1}}{BH - bh}}$$

TABLE 1.6 *(Continued)*
Section Property Tables for Various Sections

	Circle	Hollow Circle
	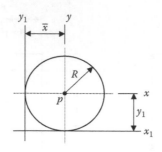	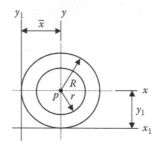

Area	$0.7854\,D^2 = \pi R^2 \quad (D = 2R)$	$\pi\!\left(R^2 - r^2\right)$
Centroid	$\bar{x} = \bar{y} = R$	$\bar{x} = \bar{y} = R$
Moment of Inertia	$I_x = I_y = \dfrac{\pi}{4}R^4$ $I_{x1} = I_{y1} = \dfrac{5\cdot\pi}{4}R^4$ $I_p = \dfrac{\pi}{2}R^4$	$I_x = I_y = \dfrac{\pi}{4}\!\left(R^4 - r^4\right)$ $I_{x1} = I_{y1} = \dfrac{\pi\!\left(5R^4 - 4R^2 r^2 - r^4\right)}{4}$ $I_p = \dfrac{\pi\!\left(R^4 - r^4\right)}{2}$
Radius of Gyration	$k_x = k_y = \dfrac{R}{2}$ $k_{x1} = k_{y1} = 1.118\cdot R$ $k_p = 0.7072\cdot R$	$k_x = k_y = \dfrac{\sqrt{R^2 + r^2}}{2}$ $k_{x1} = k_{y1} = \dfrac{\sqrt{5R^2 + r^2}}{2}$ $k_p = \sqrt{\dfrac{R^2 + r^2}{2}}$

TABLE 1.6 *(Continued)*

Section Property Tables for Various Sections

Semicircle	Hollow Semicircle
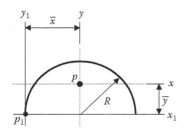	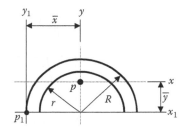

Area

$1.571R^2$

$$\frac{\pi\left(R^2 - r^2\right)}{2}$$

Centroid

$\bar{x} = R \quad \bar{y} = 0.4244\,R$

$$\bar{x} = R \quad \bar{y} = \frac{3}{4\pi}\left(\frac{R^3 - r^3}{R^2 - r^2}\right)$$

Moment of Inertia

$I_x = 0.1097\,R^4$

$I_{x1} = 0.3927\,R^4$

$I_y = 0.3927\,R^4$

$I_{y1} = 1.9637\,R^4$

$I_p = 0.5024\,R^4$

$I_{p1} = 2.3564\,R^4$

$$I_x = \frac{\pi}{8}\left(R^4 - r^4\right) - \frac{\pi\left(R^4 - r^4\right)}{2}\bar{y}^2$$

$$I_{x1} = I_y = \frac{\pi}{8}\left(R_4 - r_4\right)$$

$$I_{y1} = \frac{\pi\left(R^2 - r^2\right)\left(5R^2 + r^2\right)}{8}$$

$$I_p = I_x + I_y, \quad I_{p1} = I_{x1} + I_{y1}$$

Radius of Gyration

$k_x = 0.264\,R$

$k_{x1} = 0.5\,R$

$k_y = 0.5\,R$

$k_{y1} = 1.118\,R$

$k_p = 0.566\,R$

$k_{p1} = 1.225\,R$

$$k_x = \sqrt{\frac{2_{Ix}}{\pi\left(R^2 - r^2\right)}} \qquad k_{y1} = \sqrt{\frac{2I_{y1}}{\pi\left(R^2 - r^2\right)}}$$

$$k_{x1} = \sqrt{\frac{R^2 + r^2}{2}} \qquad k_p = \sqrt{\frac{2I_p}{\pi\left(R^2 - r^2\right)}}$$

$$k_y = \sqrt{\frac{R^2 + r^2}{2}} \qquad k_{p1} = \sqrt{\frac{2I_p}{\pi\left(R^2 - r^2\right)}}$$

TABLE 1.6 *(Continued)*
Section Property Tables for Various Sections

Rhombus	**Parallelogram**
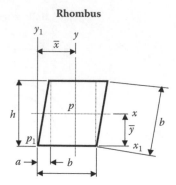	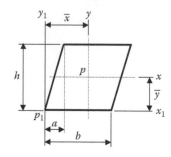

Area

$$bh \qquad\qquad\qquad bh$$

Centroid

$$\bar{x} = \frac{1}{2}(a+b) \quad \bar{y} = \frac{1}{2}h \qquad\qquad \bar{x} = \frac{1}{2}(a+b) \quad \bar{y} = \frac{1}{2}h$$

Moment of Inertia

Rhombus:

$$I_x = \frac{1}{12}bh^3$$

$$I_{x1} = \frac{1}{3}bh^3$$

$$I_y = \frac{1}{12}bh(a^2 + b^2)$$

$$I_{y1} = \frac{1}{6}bh(2a^2 + 2b^2 + 3ab)$$

$$I_p = \frac{1}{6}b^3h$$

$$I_{p1} = \frac{1}{6}b^2h(3a + 4b)$$

Parallelogram:

$$I_{x1} = \frac{1}{3}bh^3$$

$$I_y = \frac{1}{12}bh(a^2 + b^2)$$

$$I_{y1} = \frac{1}{6}bh(2a^2 + 2b^2 + 3ab)$$

$$I_p = \frac{1}{12}bh(a^2 + b^2 + h^2)$$

$$I_{p1} = \frac{1}{6}bh(2a^2 + 2b^2 + 3a^b + 2h^2)$$

Radius of Gyration

Rhombus:

$$k_x = 0.289h$$

$$k_{x1} = 0.577h$$

$$k_y = 0.289\sqrt{(a^2 + b^2)}$$

$$k_{y1} = 0.408\sqrt{(2a^2 + 2b^2 + 3ab)}$$

$$k_p = 0.408b$$

$$k_{p1} = 0.408b\sqrt{b(3a + 4b)}$$

Parallelogram:

$$k_x = 0.289h$$

$$k_{x1} = 0.577h$$

$$k_y = 0.289\sqrt{(a^2 + b^2)}$$

$$k_{y1} = 0.408\sqrt{(2a^2 + 2b^2 + 3ab)}$$

$$k_p = 0.289\sqrt{(a^2 + b^2 + h^2)}$$

$$k_{p1} = 0.408\sqrt{(2a^2 + 2b^2 + 3ab + 2h^2)}$$

TABLE 1.6 *(Continued)*
Section Property Tables for Various Sections

Right Angle Trapezoid	Isosocles Trapezoid
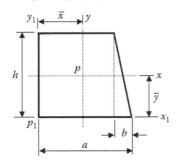	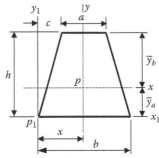

Area

$$0.5 \cdot h \cdot (2a + b)$$

$$0.5 \cdot h \cdot (a + b)$$

Centroid

$$\bar{x} = \frac{3a^2 + 3ab + b^2}{3(2a + b)} \quad \bar{y} = \frac{h(3a + b)}{3(2a + b)}$$

$$\bar{x} = \frac{b}{2} \quad \bar{y}_a = \frac{h(b + 2a)}{3(b + a)} \quad \bar{y}_b = \frac{h(a + 2b)}{3(a + b)}$$

Moment of Inertia

$$I_x = \frac{h^3 (6a^2 + 6ab + b^2)}{36(2a + b)}$$

$$I_{x1} = \frac{1}{12} h^3 (4a + b)$$

$$I_y = I_{y1} = \frac{h(3a^2 + 3ab) + b^2 (4a + b)^2}{18(2a + b)}$$

$$I_{y1} = \frac{1}{12} h \left[2a^2 (2a + 3b) + b^2 (4a + b) \right]$$

$$I_p = I_x = I_y$$

$$I_{p1} = I_{x1} = I_{y1}$$

$$I_x = \frac{h^3 (a^2 + 4ab + b^2)}{36(a + b)}$$

$$I_{x1} = \frac{h^3 (a + b)(2a + c)}{12(a + c)}$$

$$I_y = \frac{h(a + b)(a^2 + b^2)}{48}$$

$$I_{y1} = \frac{h(a + b)(a^2 + b^2)}{48}$$

$$I_p = I_x = I_y$$

$$I_{p1} = I_{x1} = I_{y1}$$

Radius of Gyration

$$k_x = \frac{0.236 h \sqrt{6a^2 + 6ab + b^2}}{2a + b}$$

$$k_{x1} = \sqrt{\frac{4a + b}{6(2a + b)}}, \quad k_y = \sqrt{\frac{2Iy}{h(2a + b)}}$$

$$k_{y1} = \sqrt{\frac{2a^2 (2a + 3b) + b^2 (4a + b)}{6(2a + b)}}$$

$$k_p = \sqrt{\frac{2I_p}{h(2a + b)}}, \quad k_{p1} = \sqrt{\frac{2I_{p1}}{h(2a + b)}}$$

$$k_x = \frac{h \sqrt{(b^2 + 4ab + a^2)}}{6(a + b)}$$

$$k_{x1} = \sqrt{\frac{h^2 (2a + c)}{a + c}},$$

$$k_y = \sqrt{\frac{a^2 + b^2}{24}} \quad k_{y1} = \sqrt{\frac{a^2 + 7b^2}{24}}$$

$$k_p = \sqrt{\frac{2I_p}{h(a + b)}}, \quad k_{p1} = \sqrt{\frac{2I_{p1}}{h(a + b)}}$$

TABLE 1.6 *(Continued)*
Section Property Tables for Various Sections

Oblique Trapezoid	Right Triangle

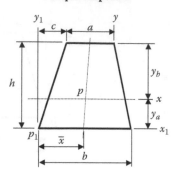 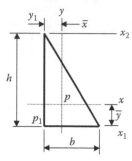

	Oblique Trapezoid	Right Triangle
Area	$\dfrac{1}{2}h(a+b)$	$\dfrac{1}{2}bh$
Centroid	$\bar{x} = \dfrac{1}{3}\left[a+b+c-\dfrac{a(b-c)}{a+b}\right]$ $\bar{y} = \dfrac{h(b+2a)}{3(a+b)}$ $\quad$ $\bar{y}_b = \dfrac{h(2b+c)}{3(a+b)}$	$\bar{x} = \dfrac{1}{2}b, \quad \bar{y} = \dfrac{1}{3}h$
Moment of Inertia	$I_x = \dfrac{h^3(a^2+4ab+b^2)}{36(a+b)}$ $I_{x1} = \dfrac{1}{12}h^3(b+3a)$	$I_x = \dfrac{1}{36}bh^3, \quad I_{x1} = \dfrac{1}{12}bh^3$ $I_{x2} = \dfrac{1}{4}bh^3, \quad I_y = \dfrac{1}{36}b^3h$ $I_{y1} = \dfrac{1}{12}b^3h$ $I_p = \dfrac{1}{36}bh(b^2+h^2)$ $I_{p1} = \dfrac{1}{12}bh(b^2+h^2)$
Radius of Gyration	$k_x = h\sqrt{\dfrac{2(a^2+4ab+b^2)}{6(a+b)}}$ $k_{x1} = h\sqrt{\dfrac{3a+b}{6(a+b)}}$	$k_x = 0.236h, \quad k_{x1} = 0.408h$ $k_{x2} = 0.707h, \quad k_y = 0.236h$ $k_{y1} = 0.408h$ $k_p = 0.236\sqrt{b^2+h^2}$ $k_{p1} = 0.408\sqrt{b^2+h^2}$

TABLE 1.6 *(Continued)*
Section Property Tables for Various Sections

Equilateral Triangle	Isosocles Triangle
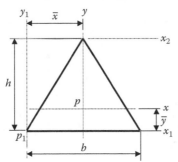	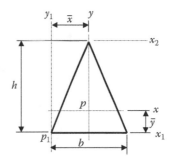

Area	$\dfrac{1}{2}bh$	$\dfrac{1}{2}bh$
Centroid	$\bar{x} = \dfrac{1}{2}b, \quad \bar{y} = \dfrac{1}{3}h$	$\bar{x} = \dfrac{1}{2}b, \quad \bar{y} = \dfrac{1}{3}h$

Moment of Inertia

Equilateral Triangle:

$$I_x = \frac{1}{36}bh^3, \quad I_{x1} = \frac{1}{12}bh^3$$

$$I_{x2} = \frac{1}{4}bh^3, \quad I_y = \frac{1}{48}hb^3$$

$$I_{y1} = \frac{7}{48}b^3h$$

$$I_p = \frac{1}{24}b^3h, \quad I_{p1} = \frac{5}{24}b^3h$$

Isosocles Triangle:

$$I_x = \frac{1}{36}bh^3, \quad I_{x1} = \frac{1}{12}bh^3$$

$$I_{x2} = \frac{1}{4}bh^3, \quad I_y = \frac{1}{48}b^3h$$

$$I_{y1} = \frac{7}{48}b^3h$$

$$I_p = \frac{1}{144}\left(4bh^3 + 3b^3h\right)$$

$$I_{p1} = \frac{1}{48}\left(4bh^3 + 7b^3h\right)$$

Radius of Gyration

Equilateral Triangle:

$$k_x = 0.236h, \quad k_{x1} = 0.408h$$
$$k_{x2} = 0.707h, \quad k_y = 0.204b$$
$$k_{y1} = 0.504b$$
$$k_p = 0.289b, \quad k_{p1} = 0.646b$$

Isosocles Triangle:

$$k_x = 0.236h, \quad k_{x1} = 0.408h$$
$$k_{x2} = 0.707h, \quad k_y = 0.204b$$
$$k_{y1} = 0.504b$$
$$k_p = 0.118\sqrt{4h^2 + 3b^2}$$
$$k_{p1} = 0.204\sqrt{4h^2 + 7b^2}$$

TABLE 1.6 *(Continued)*
Section Property Tables for Various Sections

Equilateral Triangle	Isosocles Triangle
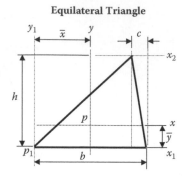	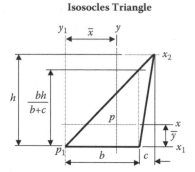

Area	$\dfrac{1}{2}bh$	$\dfrac{1}{2}bh$
Centroid	$\bar{x} = \dfrac{1}{3}(2b-c), \quad \bar{y} = \dfrac{1}{3}h$	$\bar{x} = \dfrac{1}{3}(2b-c), \quad \bar{y} = \dfrac{1}{3}h$
Moment of Inertia	$I_x = \dfrac{1}{36}bh^3, \quad I_{x1} = \dfrac{1}{12}bh^3$ $I_{x2} = \dfrac{1}{4}bh^3$ $I_y = \dfrac{1}{36}bh(b^2 + c^2 - bc)$ $I_p = \dfrac{1}{36}bh(h^2 + b^2 + c^2 - bc)$	$I_x = \dfrac{1}{36}bh^3, \quad I_{x1} = \dfrac{1}{12}bh^3$ $I_{x2} = \dfrac{1}{4}bh^3$ $I_y = \dfrac{1}{36}bh(b^2 + bc + c^2)$ $I_{y1} = \dfrac{1}{12}bh(b^2 + 3bc + 3c^2)$ $I_p = \dfrac{1}{36}bh(h^2 + b^2 + bc + c^2)$
Radius of Gyration	$k_x = 0.236h, \quad k_{x1} = 0.408h$ $k_{x2} = 0.707h,$ $k_y = 0.236\sqrt{b^2 + c^2 - bc}$ $k_p = 0.236\sqrt{h^2 + b^2 + c^2 - bc}$	$k_x = 0.236h, \quad k_{x1} = 0.408h$ $k_{x2} = 0.707h,$ $k_y = 0.236\sqrt{b^2 + bc + c^2}$ $k_{y1} = 0.408\sqrt{b^2 + 3bc + c^2}$ $k_p = 0.236\sqrt{h^2 + b^2 + bc + c}$

TABLE 1.6 *(Continued)*
Section Property Tables for Various Sections

Ellipse	Hollow Ellipse
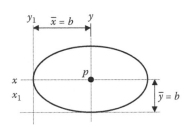	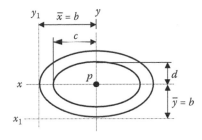

Area

$$\pi ab \qquad\qquad \pi(ab - cd)$$

Centroid

$$\bar{x} = a, \quad \bar{y} = b \qquad\qquad \bar{x} = a, \quad \bar{y} = b$$

Moment of Inertia

$$I_x = \frac{1}{4}\pi ab^3 \qquad = 0.7854\,ab^3$$

$$I_{x1} = 1.25\,\pi ab^3 \qquad = 3.927\,ab^3$$

$$I_y = \frac{1}{4}\pi a^3 b \qquad = 0.7854\,a^3 b$$

$$I_{y1} = 1.25\,\pi a^3 b \qquad = 3.927\,a^3 b$$

$$I_p = \frac{1}{4}\pi ab\left(a^2 + b^2\right)$$

$$I_x = \frac{\pi}{4}\left(ab^3 - cd^3\right), \quad I_y = \frac{\pi}{4}\left(a^3 b - c^3 d\right)$$

$$I_{x1} = \frac{\pi}{4}\left(ab^3 - cd^3\right) + \pi(ab - cd)(b)^2$$

$$I_{y1} = \frac{\pi}{4}\left(a^3 b - c^3 d\right) + \pi(ab - cd)(a)^2$$

$$I_p = I_x + I_y$$

Radius of Gyration

$$k_x = \frac{1}{2}b, \quad k_{x1} = 1.118\,b$$

$$k_y = \frac{1}{2}a, \quad k_{y1} = 1.118\,a$$

$$k_p = \frac{1}{2}\sqrt{a^2 + b^2}$$

$$k_x = \sqrt{\frac{ab^3 - cd^3}{4(ab - cd)}}, \quad k_{x1} = \sqrt{\frac{I_{x1}}{\pi(ab - cd)}}$$

$$k_y = \sqrt{\frac{a^3 b - c^3 d}{4(ab - cd)}}, \quad k_{y1} = \sqrt{\frac{I_{y1}}{\pi(ab - cd)}}$$

$$k_p = \sqrt{\frac{I_p}{\pi(ab - cd)}}$$

TABLE 1.6 *(Continued)*
Section Property Tables for Various Sections

Semi Ellipse	Hollow Semi-Ellipse
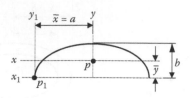	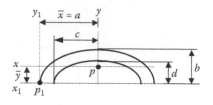

Area

$$\frac{1}{2}\pi ab$$

$$\frac{1}{2}\pi(ab - cd)$$

Centroid

$$\bar{x} = a, \quad \bar{y} = 0.424b$$

$$\bar{x} = a, \quad \bar{y} = \frac{4}{3\pi}\left[\frac{ab^2 - cd^2}{ab - cd}\right]$$

Moment of Inertia

$$I_x = 0.1098\,ab^3, \quad I_{x1} = 0.3927\,ab^3$$

$$I_y = 0.3927\,a^3b, \quad I_{y1} = 1.9635\,a^3b$$

$$I_p = ab\left(0.3927\,a^2 + 0.1098\,b^2\right)$$

$$I_{p1} = ab\left(1.9635\,a^2 + 0.3927\,b^2\right)$$

$$I_x = \frac{\pi}{8}\left(ab^3 - cd^3\right) - \frac{\pi(ab - cd)}{2}\left[\frac{4}{3\pi}\frac{ab^2 - cd^2}{ab - cd}\right]^2$$

$$I_{x1} = \frac{1}{8}\pi\left(ab^3 - cd^3\right), \quad I_y = \frac{1}{8}\pi\left(a^3b - c^3d\right)$$

$$I_{y1} = \frac{1}{8}\pi\left(a^3b - c^3d\right) + \frac{1}{2}\pi a^2(ab - cd)$$

$$I_p = I_x, \quad I_{p1} = I_{x1} + I_{y1}$$

Radius of Gyration

$$k_x = 0.2646b, \quad k_{x1} = \frac{1}{2}b$$

$$k_y = \frac{1}{2}a, \quad k_{x1} = 1.118a$$

$$k_p = \sqrt{0.25\,a^2 + 0.06987\,b^2}$$

$$k_{p1} = \sqrt{\frac{2I_{p1}}{\pi ab}}$$

$$k_x = \sqrt{\frac{2I_x}{\pi(ab - cd)}}, \quad k_{x1} = \sqrt{\frac{ab^3 - cd^3}{4(ab - cd)}}$$

$$k_y = \sqrt{\frac{2I_y}{\pi(ab - cd)}}, \quad k_{y1} = \sqrt{\frac{2I_{y1}}{\pi(ab - cd)}}$$

$$k_p = \sqrt{\frac{2I_p}{\pi(ab - cd)}}, \quad k_{p1} = \sqrt{\frac{2I_{p1}}{\pi(ab - cd)}}$$

TABLE 1.6 *(Continued)*
Section Property Tables for Various Sections

Parabolic Segment

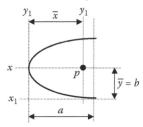

Parabolic Half-Segment

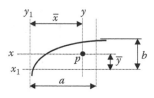

	Parabolic Segment	Parabolic Half-Segment
Area	$\dfrac{4}{3}ab$	$\dfrac{2}{3}ab$
Centroid	$\bar{x}=0.6a, \quad \bar{y}=b$	$\bar{x}=0.6a, \quad \bar{y}=0.375b$
Moment of Inertia	$I_x=0.2667\,ab^3$ $I_{x1}=1.6\,ab^3$ $I_y=0.0914\,a^3b$ $I_{y1}=0.5714\,a^3b$ $I_p=I_x+I_y$	$I_x=0.0395\,ab^3$ $I_{x1}=0.1333ab^3$ $I_y=0.0457\,a^3b$ $I_{y1}=0.2857\,a^3b$ $I_p=I_x+I_y$
Radius of Gyration	$k_x=0.4472b, \quad k_{x1}=1.095b$ $k_y=0.2619a, \quad k_{y1}=0.6546a$ $I_p=\sqrt{\dfrac{3I_p}{4ab}}$	$k_x=0.2437b, \quad k_{x1}=0.4472b$ $k_y=0.2619a, \quad k_{y1}=0.6546a$ $I_p=\sqrt{\dfrac{3I_p}{2ab}}$

TABLE 1.6 *(Continued)*
Section Property Tables for Various Sections

Regular Hexagon	Regular Octagon
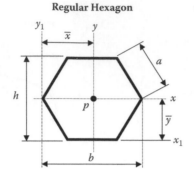	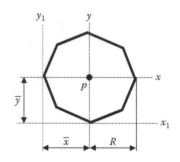

	Regular Hexagon	Regular Octagon
Area	$0.866\,h^2 = 2.598\,a^2$	$2.8286\,R^2$
Centroid	$\bar{x} = \dfrac{1}{2}b = a, \quad \bar{y} = \dfrac{1}{2}h$	$\bar{x} = \bar{y} = R$
Moment of Inertia	$I_x = I_y = 0.0601\,h^4$ $I_{x1} = 0.2766\,h^4$ $I_{y1} = 0.3488\,h^4$ $I_p = 0.1203\,h^4$	$I_x = I_y = 0.6381\,R^4$ $I_{x1} = I_{y1} = 3.4667\,R^4$ $I_p = 1.2763\,R^4$
Radius of Gyration	$k_x = k_y = 0.2635\,h = 0.4564\,a$ $k_{x1} = 0.5651\,h$ $k_{y1} = 0.6346\,h$ $k_p = 0.3727\,h$	$k_x = k_y = 0.4748\,R$ $k_{x1} = k_{y1} = 1.1072\,R$ $k_p = 0.6717\,R$

TABLE 1.6 *(Continued)*
Section Property Tables for Various Sections

Elliptical Complement

Circular Complement

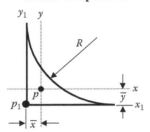

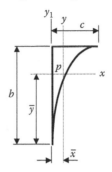

	Circular Complement	Elliptical Complement
Area	$0.2146R^2$	$0.246bc = \left(1 - \frac{1}{4}\pi\right)bc$
Centroid	$\bar{x} = \bar{y} = R$	$\bar{x} = 0.2236c, \quad \bar{y} = \dfrac{b}{1.288}$
Moment of Inertia	$I_x = I_y = 00075R^4$ $I_{x1} = I_{y1} = 0.0182R^4$ $I_p = 0.0150R^4$ $I_{p1} = 0.0365R^4$	$I_x = 0.7545b^3c$ $I_y = 0.007545bc^3$ $I_p = 0.007545bc\left(b^2 + c^2\right)$
Radius of Gyration	$k_x = k_y = 0.187R$ $k_{x1} = k_{y1} = 0.292R$ $k_p = 0.265R$ $k_{p1} = 0.412R$	$k_x = 0.18751b$ $k_y = 0.18751c$ $k_p = \sqrt{0.03516\left(b^2 + c^2\right)}$

TABLE 1.6 *(Continued)*
Section Property Tables for Various Sections

Parabola Fillet in Right Angle	Circular Section
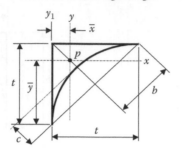	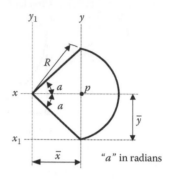
	"a" in radians

Area

$$\frac{1}{6}t^2 \qquad\qquad\qquad R^2 a$$

Centroid

$$\bar{x} = \frac{1}{5}t, \quad b = \frac{1}{\sqrt{2}}t$$

$$\bar{y} = \frac{4}{5}t, \quad c = \frac{1}{2\sqrt{2}}t$$

$$\bar{x} = \frac{2}{3}\left[\frac{R\sin a}{a}\right], \quad \bar{y} = R\sin a$$

Moment of Inertia

$$I_x = I_y = 0.00524\, t^4$$

$$I_p = 0.0105\, t^4$$

$$I_x = \frac{1}{4}R^4\left(a - \sin a \cos a\right)$$

$$I_{x1} = \frac{1}{4}R^4\left(a - \sin a \cos a\right) + R^4 a \sin^2 a$$

$$I_y = \frac{1}{4}R^4\left(a - \frac{16}{9a}\sin^2 a + \frac{1}{2}\sin^2 a\right)$$

$$I_{y1} = \frac{1}{4}R^4\left(a + \sin a \cos a\right)$$

$$I_p = \frac{1}{4}R^4\left(2a - \frac{16}{9a}\sin^2 a\right)$$

Radius of Gyration

$$k_x = k_y = 0.173t$$

$$k_p = 2.449t$$

$$k_x = \frac{1}{2}R\sqrt{1 - \frac{1}{a}\sin a \cos a}, \quad k_{x1} = \sqrt{\frac{I_{x1}}{R^2 a}}$$

$$k_y = \frac{1}{2}R\sqrt{1 + \frac{1}{a}\sin a \cos a - \frac{16}{9a^2}\sin^2 a}$$

$$k_{y1} = \frac{1}{2}R\sqrt{1 + \frac{1}{a}\sin a \cos a}$$

$$k_p = \frac{1}{2}R\sqrt{2 - \frac{16}{9a^2}\sin^2 a}$$

TABLE 1.6 *(Continued)*
Section Property Tables for Various Sections

Hollow Circular Section

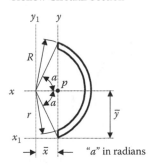

"a" in radians

Area

$$\left(R^2 - r^2\right)a$$

Centroid

$$\bar{x} = \frac{2\sin a\left(R^3 - r^3\right)}{3a\left(R^2 - r^2\right)}, \quad \bar{y} = R\sin a$$

Moment of Inertia

$$I_x = \frac{1}{4}a\left(R^4 - r^4\right)\left(1 - \frac{1}{a}\sin a\cos a\right)$$

$$I_{x1} = \frac{1}{4}a\left(R^4 - r^4\right)\left(1 - \frac{1}{a}\sin a\cos a\right) + a\left(R^4 - r^4 R^2\right)\sin^2 a$$

$$I_y = \frac{1}{4}a\left(R^4 - r^4\right)\left(1 + \frac{1}{a}\sin a\cos a\right) \quad \frac{1}{a\left(R^2 - r^2\right)}\left[\frac{2}{3}\sin a\left(R^3 - r^3\right)\right]^2$$

$$I_{y1} = \frac{1}{4}a\left(R^4 - r^4\right)\left(1 + \frac{1}{a}\sin a\cos a\right)$$

$$I_p = I_x + I_y$$

Radius of Gyration

$$k_x = \frac{1}{2}\sqrt{\left(R^2 + r^2\right)\left(1 - \frac{1}{a}\sin a\cos a\right)}$$

$$k_{x1} = \sqrt{\frac{I_{x1}}{\left(R^2 - r^2\right)a}}, \quad k_y = \sqrt{\frac{I_y}{\left(R^2 - r^2\right)a}}$$

$$k_{y1} = \frac{1}{2}\sqrt{\left(R^2 + r^2\right)\left(1 + \frac{1}{a}\sin a\cos a\right)}$$

$$k_p = \sqrt{\frac{I_p}{\left(R^2 - r^2\right)a}}$$

TABLE 1.6 *(Continued)*
Section Property Tables for Various Sections

Circular Segment

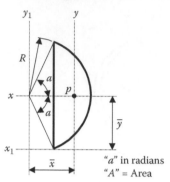

"*a*" in radians
"*A*" = Area

I Section

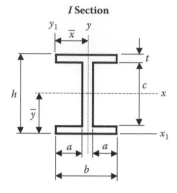

	Circular Segment	I Section
Area	$\dfrac{1}{2}R2(2a - \sin 2a)$	$bh - 2ac$
Centroid	$\bar{x} = \dfrac{4R\sin 3a}{3(2a - \sin 2a)}, \quad \bar{y} = R\sin a$	$\bar{x} = \dfrac{1}{2}b$ $\bar{y} = \dfrac{1}{2}h$

Moment of Inertia

$$I_x = \frac{1}{4}AR^2\left[1 - \frac{2}{3}\left(\frac{\sin^3 a\cos a}{a - \sin a\cos a}\right)\right]$$

$$I_{x1} = 1_x + \frac{1}{2}R^4(2a - \sin 2a)(\sin^2 a)$$

$$I_y = \frac{AR^2}{4}\left[1 + \frac{2\sin^3 a\cos a}{a - \sin a\cos a}\right] - \frac{4R^6\sin^6 a}{9A}$$

$$I_y = \frac{AR^2}{4}\left[1 - \frac{2\sin^3 a\cos a}{a - \sin a\cos a}\right]$$

$$I_p = I_x + I_y$$

$$I_x = \frac{1}{12}\left(bh^3 - 2Ac^3\right)$$

$$I_{x1} = \frac{1}{3}bh^3 Ac\left(c^2 + 3h^2\right)$$

$$I_y = \frac{1}{12}t\left(2b^3 + ct^2\right)$$

Radius of Gyration

$$k_x = \sqrt{\frac{R^2}{4}\left(1 - \frac{2\sin^3 a\cos a}{3(a - \sin a\cos a)}\right)}$$

$$k_y = \sqrt{\frac{2I_y}{R^2(2a - \sin a)}}$$

$$k_p = \sqrt{\frac{2I_p}{R^2(2a - \sin 2a)}}$$

$$k_x = \sqrt{\frac{bh^3 - 2Ac^3}{12(bh - 2Ac)}}$$

TABLE 1.6 *(Continued)*
Section Property Tables for Various Sections

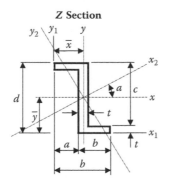

Z Section

T Section

Area	$t(d+2a)$	$ae+bd$

Centroid

$$\bar{y} = \frac{d}{2}, \quad \bar{x} = \frac{2b-t}{2}$$

$$y_1 = h - \frac{1}{2}\left(\frac{dh^2 + 2ce^2}{dh + 2ce}\right), \quad y_2 = h - y_1$$

Moment of Inertia

$$I_x = \frac{1}{12}\left(bd^3 - a(d-2t)^3\right)$$

$$I_{x2} = I_x \cos^2\alpha + I_y \sin^2\alpha - I_{xy}\sin 2\alpha$$

$$I_y = \frac{1}{12}\left(d(b+a)^3 - 2a^3c - 6ab^2c\right)$$

$$I_{y2} = I_y \cos^2\alpha + I_x \sin^2\alpha + I_{xy}\sin 2\alpha$$

$$\tan 2\alpha = \frac{abct}{I_x - I_y}$$

$$I_p = I_x + I_y, \quad I_{xy} = \frac{1}{2}abct$$

$$I_x = \frac{1}{3}\left[ay_2^3 - 2c(b-y_1)^3 + dy_1^3\right]$$

$$I_y = \frac{ea^3 + bd^3}{12}$$

Radius of Gyration

$$k_x = \sqrt{\frac{I_x}{t(d+2a)}}$$

$$k_y = \sqrt{\frac{I_y}{t(d+2a)}}$$

$$k_x = \sqrt{\frac{I_x}{ae+bd}}$$

$$k_y = \sqrt{\frac{ea^3 + bd^3}{12(ea+bd)}}$$

1.5.5 SECTION SHEAR CENTERS

The shear center (also known as the flexural center) can be defined as the point in a plane of the cross section through which an applied transverse load must pass for bending to occur unaccompanied by twisting.

To obtain the shear center a unit transverse load is applied to the section at the datum point.

The shear distribution, thus shear flow distribution, thus shear force distribution, is obtained for each element of the section.

Equating the torque given by the shear distribution to the torque given by the transverse load will produce the position of the shear center.

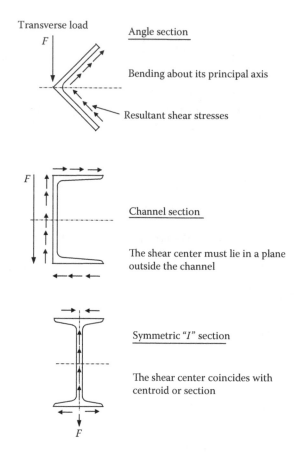

FIGURE 1.36 Shear flows around various sections.

1.5.5.1   Location of Shear Center: Open Sections

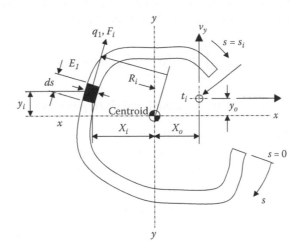

FIGURE 1.37   Location of shear center–open section.

1.5.5.2   Section Constants

$$Z_x = \frac{I_{xx}}{I_{xx} \cdot I_{yy} - I_{xy}^2}$$

$$Z_y = \frac{I_{yy}}{I_{xx} \cdot I_{yy} - I_{xy}^2}$$

$$Z_{xy} = \frac{I_{xy}}{I_{xx} \cdot I_{yy} - I_{xy}^2}$$

1.5.5.3   Shear Flow in an Element

$$q_i = \int_0^{s_i} t_i \left[\left(x_i \cdot V_y + y_i \cdot V_x \right) Z_{xy} - x_i \cdot V_x \cdot Z_x - y_i \cdot V_y \cdot Z_y \right] ds + q_o$$

where:

q_o = the value of q_i at s = o
x_i = horizontal distance from the element centroid to the section centroid
y_i = vertical distance from the element centroid to the section centroid

1.5.5.4   Shear Force in an Element

$$F_i = q_i \cdot ds$$

1.5.5.5   Torque/Moment Given by Element

$$T_i = F_i \cdot R_i$$

Centroid chosen as datum point, but it could be any convenient point.

1.5.5.6 Horizontal Location of the Shear Center

To locate the horizontal position of the shear center, we apply a vertical unit force through the unknown position of the shear center x_o, y_o, i.e., $V_y = 1$ and $V_x = 0$.

$$\therefore\ q_i = \int_0^{s_i} \left(x_i \cdot Z_{xy} - y_i\ Z_y \right) t_i \cdot ds + q_o$$

Shear forces and total torque $\Sigma\ T_i$ are calculated as defined previously.

1.5.5.7 Vertical Location of the Shear Center

To locate the vertical position of the shear center, a horizontal unit force is applied through the unknown position of the shear center x_o, y_o, i.e., $V_y = 0$ and $V_x = 1$.

$$\therefore\ q_i = \int_0^{s_i} \left(y_i \cdot Z_{xy} - x_i\ Z_x \right) t_i \cdot ds + q_o$$

Note:

If a beam cross-section has two axes of symmetry, the shear center will be located at the centroid of the section. If the beam has one axis of symmetry, the shear center will be located at one axis of symmetry.

1.5.5.8 Shear Center of a Curved Web

$$q\theta = \int_0^\theta -y \cdot \left(\frac{1}{I_{xx}} \right) t \cdot ds$$
$$y = R \cdot \cos\theta \qquad ds = R\ d\theta$$
$$q\theta = \frac{t}{I_{xx}} \int_0^\theta R \cos\theta \cdot R\ d\theta$$
$$q\theta = \frac{t}{I_{xx}} R^2 \sin\theta$$

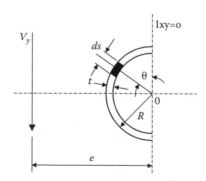

FIGURE 1.38   Shear center for a curved web.

1.5.5.9 Horizontal Location

$$= q \cdot ds \cdot R$$

$$= q \cdot Rds \cdot R$$

$$= \frac{R^4 \cdot t}{I_x} \sin\theta \cdot d\theta$$

Moment of force of element about 0.

Total moment:

$$= \frac{R^4 t}{I_{xx}} \int_0^\pi \sin\theta \, d\theta$$

$$= \frac{R^4 t}{I_{xx}} [-\cos\theta]_0^\pi$$

$$= \frac{2R^4 t}{I_{xx}}$$

Vertical component of force of element:

$$= q \cdot ds \cdot \sin\theta$$

Therefore total vertical force:

$$= \frac{R^3 t}{I_{xx}} \int_0^\pi \sin^2\theta \, d\theta$$

$$= \frac{R^3 t \cdot \pi}{2 \, I_{xx}}$$

Taking moments about 0,

$$\frac{R^3 t \cdot \pi}{2 \, I_{xx}} \cdot e = \frac{2 \, R^4 t}{I_{xx}}$$

Horizontal location:

$$e = \frac{4 \, R}{\pi}$$

Alternatively for a solid section:

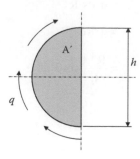

FIGURE 1.39   Shear flow around a semicircle.

where
   q = web shear flow
   A = enclosed area
   h = web depth

Vertical equilibrium:

$$qh = V_y$$

Moment equilibrium:

$$2\,Aq = V_y \cdot e$$

$$e = \frac{2A}{h}$$

TABLE 1.7
Shear Center Positions for Various Sections

Section   **Position of Shear Center**

1. Open Thin-Walled Circular Tube

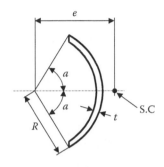

$$e = 2R\left[\frac{\sin\alpha - \alpha\cos\alpha}{\alpha - \sin\alpha\cos\alpha}\right]$$

For $\alpha = \dfrac{\pi}{2}$ $e = \dfrac{4R}{\pi}$

For $\alpha = \pi$ $e = 2R$

where α is in radians

2. Semicircular Area

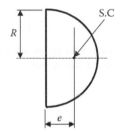

$$e = \left(\frac{8}{15\pi}\left(\frac{3+4\nu}{1+\nu}\right)\right)R$$

where ν = Poisson's Ratio

3. Channel—Equal Flanges and Uniform Thickness

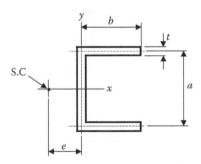

$$e = a\frac{H_{xy}}{I_x}$$

where H_{xy} is the product of inertia
for the area above the x axis

If t is uniform and small:

$$e = \frac{a^2 b^2 t}{4 I_x} = b\left(\frac{3\left(\dfrac{b}{a}\right)}{1 + 6\left(\dfrac{b}{a}\right)}\right)$$

I_x is the moment of inertia of
whole section about the "x" x axis

TABLE 1.7 *(Coninued)*
Shear Center Positions for Various Sections

Section	Position of Shear Center

4. Section—Equal Flanges

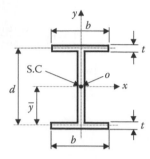

Shear center at centroid
$$x_o = 0$$
$$y_o = 0$$

5. Section with Unequal Flanges and Uniform Thin Web

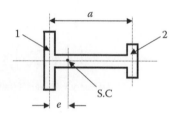

$$e = a\left(\frac{I_2}{I_1 + I_2}\right)$$

where I_1 and I_2 are the moments of inertia of flanges 1 and 2 about the "x" axis

6. Section—Unequal Flanges

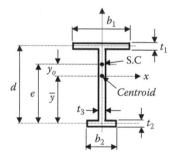

$$x_o = 0$$
$$y_o = e - \bar{y}$$
$$e = \frac{dt_1 b_1^3}{t_1 b_1^3 + t_2 b_2^3}$$

FURTHER READING

Constrado. 1983. *Steel Designers' Manual.* 4th ed. UK.
Young, Warren C. *Roark's Formulas for Stress and Strain.* 1989. 6th ed. New York: McGraw-Hill.

2 Torsion of Solid Sections

2.1 INTRODUCTION

If a shaft is subjected to a pure torque, i.e., without any bending, buckling, or axial thrust, every cross section is in a state of pure shear. The shearing stress induced in the shaft produces a moment of resistance, equal and opposite to the applied torque.

List of Symbols

$$
\begin{aligned}
T &= \text{Applied torsional moment} \\
L &= \text{Length of member} \\
G &= \text{Modulus of rigidity} \\
Ip &= \text{Polar moment of inertia of section} \\
J &= \text{Torsion constant of section} \\
\tau &= \text{Applied shear stress} \\
q &= \text{Applied shear flow} \\
\theta &= \text{Angle of twist} \\
A &= \text{Area of cross section}
\end{aligned}
$$

2.2 BASIC THEORY

In the theory of twisting, which relates the shear stress and angle of twist to the applied torque, it is assumed that:

1. The material is homogeneous, elastic, and obeys Hooke's law; i.e., the shear stress at any point is proportional to the shear strain at that point.
2. Stresses do not exceed the limits of proportionality.
3. Radial lines remain radial after twisting.
4. The plane cross section remains plane after twisting. (For noncircular shafts, the assumption that the plane sections remain plane after twisting is not justified and this theory ceases to apply.)

From the third assumption above, it follows that the strain (and hence the stress) is directly proportional to the radius. Thus, if the shear stress at the surface of the shaft is τ, then the stress τ' on an element da at a distance x from the axis (see Figure 2.1).

$$\tau' = \frac{x}{r}\tau$$

$$\therefore \text{ shear force on element} = \frac{x}{r}\tau\, da$$

$$\therefore \text{ moment of force about O} = \frac{x}{r}\tau\, da\, x$$

$$\therefore \text{total moment of resistance} = \frac{\tau}{r}\int x^2 da$$

$$= \frac{\tau}{r}J$$

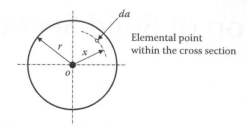

FIGURE 2.1   Cross section of a circular shaft subject to pure torsion.

This is equal to the applied torque T, i.e.,

$$T = \frac{\tau}{r} J$$

or

$$\frac{T}{J} = \frac{\tau}{r}$$

(2.1)

This formula gives the shear stress at the surface in terms of T and J, but the stress at any other radius can be readily found since it is proportional to the radius at that point.

2.3   MODULUS OF SECTION

The maximum shear stress in a shaft is given by

$$\tau = \frac{T}{J} r$$

$$= \frac{T}{J/r}$$

The quantity J/r is known as the Modulus of Section and is denoted by Z.

Thus

$$\tau = \frac{T}{Z}$$

(2.2)

For a solid shaft

$$Z = \frac{\dfrac{\pi d^4}{32}}{\dfrac{d}{2}}$$

(2.3)

$$= \frac{\pi}{16} d^3$$

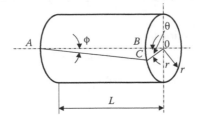

FIGURE 2.2 Angle of twist.

For a hollow shaft

$$Z = \frac{\dfrac{\pi\left(D^4 - d^4\right)}{32}}{\dfrac{D}{2}} \tag{2.4}$$

where D and d are the outer and inner diameters, respectively.

$$Z = \frac{\pi}{16}\left(\frac{D^4 - d^4}{D}\right) \tag{2.5}$$

2.4 ANGLE OF TWIST

Due to the shear strain in the shaft, the longitudinal line AB (see Figure 2.2) will rotate to the position AC, the end A being considered fixed.

The angle BAC is the shear strain, φ, and the angle BOC is the angle of twist, θ.

$$BC = L\varphi = r\theta$$

$$\therefore \varphi = \frac{r}{L}\theta = \frac{\tau}{G} \tag{2.6}$$

$$\therefore \frac{\tau}{r} = \frac{G\theta}{L}$$

Combining this with Equation (2.1) gives the general twisting formula:

$$\frac{T}{J} = \frac{\tau}{r} = \frac{G\theta}{L} \tag{2.7}$$

2.5 PURE TORSION OF OPEN SECTIONS

2.5.1 THICK-WALLED OPEN SECTIONS

These types of sections have relatively thick rectangular elements having fillets at the junctions.

TABLE 2.1

Torsional Properties of Solid Sections

Cross-Section	Torsional Constant $J \text{ in } \theta = \dfrac{Tl}{GJ}$	Torsional Shear Stress f_s
Circle 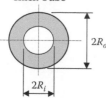 $2R$	$J = \dfrac{\pi R^4}{2}$	$\text{Max } f_s = \dfrac{2T}{\pi R^3}$ at outer surface
Thick Tube 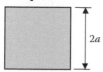 $2R_o$ $2R_i$	$J = \dfrac{\pi \left(R_o^4 - R_i^4\right)}{2}$	$\text{Max } f_s = \dfrac{2TR_o}{\pi \left(R_o^4 - R_i^4\right)}$ at outer surface
Solid Ellipse 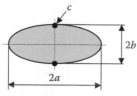 c $2b$ $2a$	$J = \dfrac{\pi \cdot a^3 \cdot b^3}{a^2 + b^2}$	$\text{Max } f_s = \dfrac{2T}{\pi \cdot a \cdot b_2}$ at c
Square $2a$	$J = 2.25 \cdot a^4$	$\text{Max } f_s = \dfrac{0.601 \cdot T}{a^3}$ At midpoint of each side
Rectangular c c	$J = ab^3 \left[\dfrac{16}{3} - 3.36 \dfrac{a}{b}\left(1 - \dfrac{b^4}{12a^4}\right)\right]$ For $a \geq b$	$\text{Max } f_s = T\left[\dfrac{3a + 1.8b}{8a^2 b^2}\right]$ At midpoint of long sides
Circular Segment 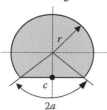 r c $2a$	$J = 2cr^4$ $h = r(1 - \cos\alpha)$ for $0 \leq \dfrac{h}{4} \leq 1.0$ $c = 0.7854 - 0.3333\dfrac{h}{r}$ $-2.6183\left(\dfrac{h}{r}\right)^2 + 4.1595\left(\dfrac{h}{r}\right)^3$ $-3.0769\left(\dfrac{h}{r}\right)^4 + 0.9299\left(\dfrac{h}{r}\right)^5$	$\text{Max } f_s = \dfrac{T\beta}{r^3}\left(\text{at } c\right)$ For $0 \leq \dfrac{h}{r} \leq 1.0$ $\beta = 0.6366 + 1.7398\dfrac{h}{r}$ $-5.4897\left(\dfrac{h}{r}\right)^2 + 14.062\left(\dfrac{h}{r}\right)^3$ $-14.51\left(\dfrac{h}{r}\right)^4 + 6.434\left(\dfrac{h}{r}\right)^5$

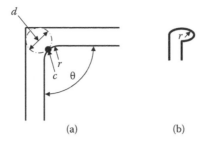

FIGURE 2.3 (a) L-shaped section. Concave curve. c = point of maximum stress; r = radius of fillet; θ = angle between elements (radians); d = diameter of largest inscribed circle; A = area of cross section. (b) Convex curve (i.e., at bulb).

Consider an L-shaped section as shown in Figure 2.3(a). At the point where the curvature is concave or re-entrant is negative, the approximate maximum stress is given by:

$$f_s = \frac{T}{J} n \qquad \text{and occurs at point } c$$

$$(f_s = \text{shear stress})$$

J for various sections is given in Tables 2.2 and 2.3.

The parameter n is defined as follows:

For the concave radii as shown in Figure 2.3(a)

where
$$n = \frac{d}{1+\left(\dfrac{\pi^2 \cdot d^4}{16A^2}\right)}\left\{1+\left[0.118\ln\left(1-\frac{d}{2r}\right)-0.238\frac{d}{2r}\right]\tanh\frac{2\theta}{\pi}\right\}$$

For convex curves (such as bulbs) as shown in Figure 2.2(b)

$$n = \frac{d}{1+\left(\dfrac{\pi^2 \cdot d^4}{16A^2}\right)}\left[1+0.15\left(\frac{\pi^2 d^4}{16A^2}-\frac{d}{2r}\right)\right]$$

2.5.2 THIN-WALLED OPEN SECTIONS

The torsion in a thin-walled open section is balanced by the shear flow, q, about the periphery of the section. Consider Figure 2.4. The stress distribution is assumed to be linear across the thickness of the section as indicated in Figure 2.4.

$$\text{Torsional Constant } J = \frac{1}{3}\sum_{i=1}^{n} b_i t_i^3$$

$$\text{Maximum shear stress } f_s = \frac{T}{J} \cdot t$$

Note that this applies to pure torsion only; the effect of warping and associated differential bending loads in the flanges will be considered separately (see Table 2.4).

TABLE 2.2

Torsional Constants for Various Sections (Part 1)

Section	Torsional Constant
	J as in $\dfrac{T\ell}{GJ}$ and in $f_s = \dfrac{T}{J}n$

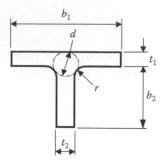

$J = K_1 + K_2 + ad^4$

if $t_1 \geq t_2$

$$a = \frac{t_2}{t_1}\left(0.15 + 0.10\frac{r}{t_1}\right)$$

if $t_2 \geq t_1$

$$a = \frac{t_1}{t_2}\left(0.15 + 0.10\frac{r}{t_1}\right)$$

for $t_2 < 2(t_1 + r)$

$$d = 2\left[t_2 + t_1 + 3r - \sqrt{2(2r + t_1)(2r + t_2)}\right]$$

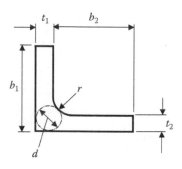

$J = K_1 + K_2 + ad^4$

if $t_1 \geq t_2$

$$a = \frac{t_2}{t_1}\left(0.07 + 0.076\frac{r}{t_1}\right)$$

if $t_2 \geq t_1$

$$d = 2\left[t_2 + t_1 + 3r - \sqrt{2(2r + t_1)(2r + t_2)}\right]$$

2.6   THIN-WALLED CLOSED SECTIONS

2.6.1   SINGLE CELL SECTIONS

The torsion of thin-walled closed sections is balanced at any section along a member by a uniform shear flow around the section. The stress is assumed to be uniform through the thickness of the section.

It will be obvious that the maximum stress will occur at the thinnest part of the section (see Figure 2.5).

$$\text{Shear flow q } (\text{load }/\text{mm}) = \frac{T}{2A} \tag{2.8}$$

$$\text{Shear stress } f_s = \frac{T}{2At} = \frac{q}{t} \tag{2.9}$$

The torsional constant J for thin walled closed sections subjected to a torque is given by:

$$J = \frac{4A^2}{\displaystyle\int_0^b \frac{d_s}{t}} = \frac{4A2}{\displaystyle\sum_{i=1}^{n} \frac{b_i}{t_i}}$$

TABLE 2.3

Torsional Constants for Various Sections (Part 2)

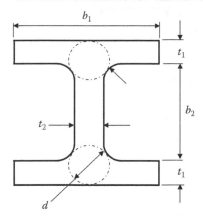

$$J = K_1 + K_2 + ad^4$$

$$a = \frac{t}{t'}\left(0.15 + 0.1\frac{r}{t_1}\right)$$

where

t = minimum of t_1 or t_2

t' = maximum of t_1 or t_2

For $t_2 < 2(t_1 + t_2)$

$$d = \frac{(t_1 + r)^2 + r \cdot t_2 + \frac{t_2^2}{4}}{(2r + t_1)}$$

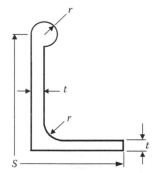

$$J = \frac{1}{3}st^3\left[1 + \left(\frac{4.14 + 4.71\left(\frac{r}{t}\right)^4}{\frac{s}{t}}\right)\right]$$

for 'n' at bulb or fillet see paragraph (2.1.6.1)

Sections composed of 3 or more rectangular elements.

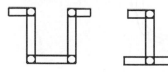

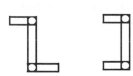

$J = \Sigma J$ of the constituent tee and angle sections:

$$= \Sigma K_1 + \Sigma K_2 + \Sigma a \cdot d^4$$

Where K_1 and K_2 are given as:

$$K_1 = b_1 t_1^3\left[\frac{1}{3} - 0.21\frac{t_1}{b_1}\left(1 - \frac{t_1^4}{12b_1^4}\right)\right]$$

$$K_2 = b_2 t_2^3\left[\frac{1}{3} - 0.105\frac{t_2}{b_2}\left(1 - \frac{t_2^4}{192b_2^4}\right)\right]$$

$$K_3 = \frac{1}{3}b_2 t_2^3$$

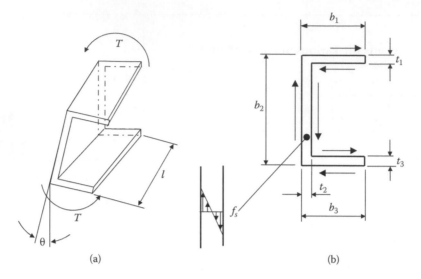

FIGURE 2.4 Torsion in a thin-walled section.

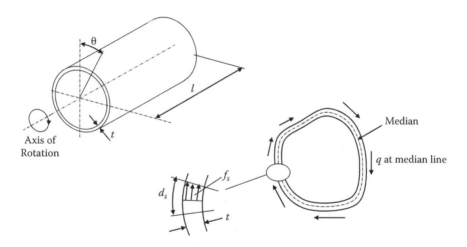

FIGURE 2.5 Torsion in a closed section. A = area enclosed by median contour; b = total length of median contour; d_s = length of differential element along median contour; t = local wall thickness.

$$b_i = \text{length of portion of the wall}$$

$$t_i = \text{thickness of portion of the wall}$$

$$\text{angle of twist } q = \frac{Tl}{GJ} \tag{2.10}$$

$$\text{rate of twist } \frac{\theta}{l} = \frac{q}{2AG} \int \frac{d_s}{t} \tag{2.11}$$

TABLE 2.4

Torsional Constants for Thin-Walled Open Sections

Cross-Section	Torsional Constant	Torsional Shear Stress
	J in $\theta = \dfrac{Tl}{GJ}$	$f_s = \dfrac{T}{J}t$

Thin Rectangular

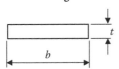

$$J = \frac{1}{3}b \cdot t^3$$

$$f_s = \frac{3T}{bt^2}$$

Thin Circular

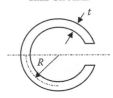

$$J = \frac{2}{3}\pi R t^3$$

$$f_s = \frac{3T}{2\pi R T_2}$$

R = Mean radius.

Any Formed Section

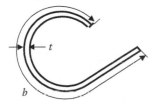

$$J = \frac{1}{3}b \cdot t^3$$

$$f_s = \frac{3T}{bt^2}$$

$$J = \frac{1}{3}\sum_{i=1}^{n} bi \cdot ti^3$$

$$f_{si} = \frac{T}{J}rt_i$$

$$f_{s\,max} = \frac{T}{J}t_{max}$$

2.7 CURVED MEMBERS

2.7.1 CURVED TORSION MEMBERS

In the previous section the maximum torsional shear stress on straight members was shown. This section will introduce an additional factor, k, to account for the effects of curvature on curved members.

If the member's radius of curvature is less than 10 times the member's depth, i.e., R/d < 10, then this factor will need to be applied. It is also assumed that the material properties will remain elastic.

TABLE 2.5

Torsional Constants for Thin-Walled Closed Sections

Cross-Section	Torsional Constant	Torsional Shear Stress
	J in $\theta = \dfrac{Tl}{GJ}$	$f_s = \dfrac{T}{2at}$

Tube

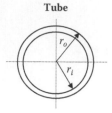

$$J = \frac{1}{2}\pi\left(r_o^4 - r_i^4\right)$$

$$f_s = \frac{2Tr_o}{\pi\left(r_o^4 - r_i^4\right)}$$

Ellipse

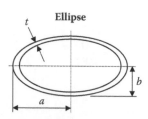

$$J = \frac{4\pi^2 t\left[\left(a-\frac{1}{2}t\right)^2\left(b-\frac{1}{2}t\right)^2\right]}{U}$$

U = length of elliptical median boundary

$$U = \pi(a+b-t)\left[1+0.258\frac{(a-b)^2}{(a+b-t)^2}\right]$$

$$f_s = \frac{2T}{2\pi t\left(a-\frac{1}{2}t\right)\left(b-\frac{1}{2}t\right)}$$

Rectangle

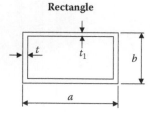

$$J = \frac{2tt_1(a-t)^2(b-t_1)^2}{at+bt_1-t^2-t_1^2}$$

$$f_{s1} = \frac{T}{2t(a-t)(b-t_1)} \quad \text{Short sides}$$

$$f_{s2} = \frac{T}{2t_1(a-t)(b-t_1)} \quad \text{Long sides}$$

General

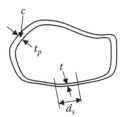

$$J = \frac{4A^2}{\int_0^b \frac{d}{t}} = \frac{4A^2}{\sum_{i=1}^n \frac{b_i}{t_i}}$$

Constant t:

$$J = 4A^2\frac{t}{b}$$

at c

$$F_{sp} = \frac{T}{At_p}$$

$$f_s = \frac{T}{2At}$$

A = area of enclosed cell; b = length of median line.

2.7.2 CIRCULAR SECTION

For a straight member:

$$f_{s_{max}} = \frac{T\cdot\left(\frac{d}{2}\right)}{J}$$

$$= \frac{16\cdot T}{\pi\cdot d^3}$$

(2.12)

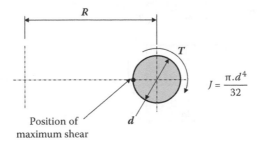

FIGURE 2.6　Torsion in a circular section.

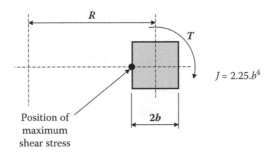

FIGURE 2.7　Torsion in a square section.

For a curved member:

$$f_{s_{max}} = k_1 \cdot \left[\frac{16 \cdot T}{\pi \cdot d^3} \right]$$

$$k_1 = 1 + \frac{5}{8} \left(\frac{d}{R} \right) + \frac{7}{32} \left(\frac{d}{R} \right)^2$$

(2.13)

2.7.3　SQUARE SECTION

For a straight member:

$$f_{s_{max}} = \frac{0.601 \cdot T}{b^3}$$

(2.14)

For a curved member:

$$f_{s_{max}} = k_2 \left[\frac{0.601 \cdot T}{b^3} \right]$$

(2.15)

$$k_2 = 1 + 1.2 \left(\frac{b}{R} \right) + 0.56 \left(\frac{b}{R} \right)^2 + 0.5 \left(\frac{b}{R} \right)^3$$

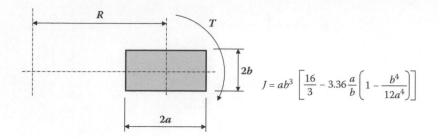

FIGURE 2.8 Torsion in a rectangular section.

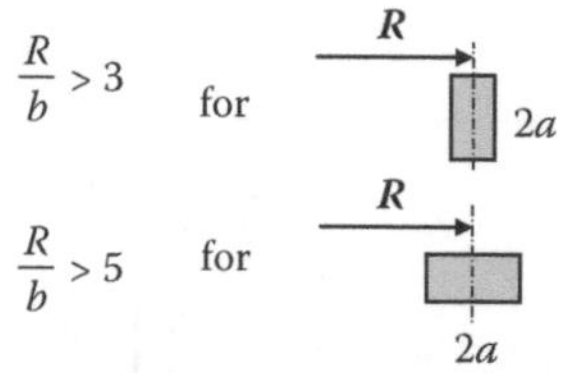

$$\frac{R}{b} > 3 \quad \text{for}$$

$$\frac{R}{b} > 5 \quad \text{for}$$

FIGURE 2.9 Ratio of R to b for the aspect ratio of the section.

2.7.4 RECTANGULAR SECTIONS

For a straight member:

$$f_{s_{max}} = T\left[\frac{3a + 1.8b}{8 \cdot a^2 \cdot b^2}\right] \tag{2.16}$$

For a curved member:

$$f_{s_{max}} = k_2 \cdot T\left[\frac{3a + 1.8b}{8 \cdot a^2 \cdot b^2}\right]$$

$$k_2 = 1 + 1.2\left(\frac{b}{R}\right) + 0.56\left(\frac{b}{R}\right)^2 + 0.5\left(\frac{b}{R}\right)^3 \tag{2.17}$$

The above formulas will apply as in Figure 2.9.

2.7.5 SPRINGS

Spring Nomenclature

α = Pitch angle of coils
n = Number of active coils in spring
δ = Axial deflection
G = Modulus of rigidity
P = Applied force
ν = Poisson's ratio

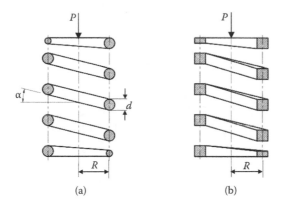

FIGURE 2.10   Spring forms. (a) Circular section; (b) square section.

The axial deflections of springs are as follows:

For circular sections:

$$\delta = \frac{64P \cdot R^3 \cdot n}{G \cdot d^4} \left[1 - \frac{3}{64} \left(\frac{d}{R} \right)^2 + \frac{3+\nu}{2(1+\nu)} (\tan a)^2 \right] \qquad (2.18)$$

For square sections:

$$\delta = \frac{2.789P \cdot R^3 \cdot n}{G \cdot b^4} \qquad \text{for } c > 3 \qquad (2.19)$$

For rectangular sections:

$$\delta = \frac{3 \cdot \pi \cdot P \cdot R^3 \cdot n}{8 \cdot G \cdot b^4} \left[\frac{a}{b} - 0.627 \left(\tanh \left(\frac{\pi b}{2a} \right) + 0.004 \right) \right]^{-1} \qquad (2.20)$$

2.8   TORSIONAL FAILURE OF TUBES

Failure in tubes due to torsion may be attributable to material failure or instability of the tube walls. The instability may be due to either:

- d/t ratio may be too large
- plastic regions existing in the wall of the tube

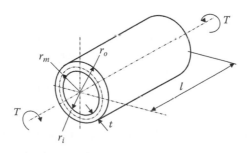

FIGURE 2.11   Notation.

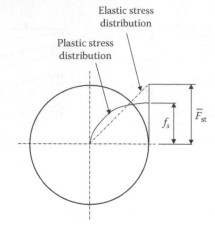

FIGURE 2.12   Elastic and plastic stress distribution.

r_m = radius to mean thickness
t = wall thickness
l = tube length
T = applied torque
$\overline{T}$ = allowable torque
J = torsional constant
f_s = applied shear stress
$\overline{f}_{sbe}$ = allowable elastic buckling stress at the mean radius (r_m)
$\overline{F}_{st}$ = allowable shear stress at r_o
$\overline{F}_{stb}$ = allowable buckling stress at the outside radius (r_o)
$\overline{F}_{stp}$ = allowable plastic stress at r_o
$\overline{F}_{su}$ = allowable ultimate shear stress for tube material

The shear stress curve is similar to that of the tensile stress curve with a $1/\sqrt{3}$ factor applied to the ordinates. Plastic torsion in tubes is treated similar to a bar in plastic bending (Figure 2.12 shows the elastic and plastic stress distribution in a solid shaft.)

The allowable shear stress, F_{st}, is also referred to as the *torsional modulus of rupture*. This is a fictitious stress acting at the surface of the tube and represents the same resisting moment as the true stress in the section.

Material property curves for the torsional modulus of rupture can be taken from MIL-HDBK 5[*] for a wide range of materials.

2.8.1   MODULUS OF RUPTURE: A THEORETICAL APPROACH

The first step to determine the modulus of rupture, F_{st}, is to calculate the stress that will lead to an instability failure, F_{stb}, and the stress to cause a material shear failure, F_{stp}.

The allowable F_{st} for a tube is the minimum of F_{stb} or F_{stp}.

The effects of any imperfections in the plastic range are considered small and can be disregarded.

2.8.1.1   Instability Failure

The elastic buckling stress (F_{stb}) is calculated by setting $E_{sec} = E$ and $Mr = 1.0$ in the following equation:

---

[*] MIL-HDBK 5 (Military Handbook 5): U.S. Government publication that covers a wide range of materials including steels, aluminium, copper, etc.

Plastic Buckling Stress (F_{sb})

$$\overline{F}_{sb} = \frac{\left[3.2074H^4\left[A\right]+0.08213H^8\left[B+C\right]\right]\cdot E_{sec}}{D} \tag{2.21}$$

where:

$$A = 37.3063K^4 + 7.0422\left(1+\frac{1}{Mr}\right)K^2 + 2$$

$$B = \frac{1}{\left(0.0276H^4 + 0.1660\left(3Mr-1\right)H^2 + 1\right)}$$

$$C = \frac{1}{\left(0.9608H^4 + 0.9802\left(3Mr-1\right)H^2 + 1\right)}$$

$$D = 10.773H\left[\frac{l}{t}\right]$$

$$H = \left[\frac{l2}{t\cdot r_m}\right]^{\frac{1}{4}}$$

$$K = \frac{1}{H}$$

$$E_{sec} = \text{Secant Modulus}$$

The modulus ratio, M_r, is the ratio of the secant to the tangent modulus at a stress level equivalent to F_{sb}; hence an iteration is required to obtain a value.

$$Mr = \frac{1+\frac{3}{7}(m)\left(\sqrt{3}\,\frac{f_s}{F_{0.7}}\right)^{m-1}}{1+\frac{3}{7}\left(\sqrt{3}\,\frac{f_s}{F_{0.7}}\right)m-1} = \frac{E_{sec}}{E_{tan}} \quad \text{at } \overline{F}_{sb} \tag{2.22}$$

E_{sec}, E_{tan}, Secant and Tangent modulii at $\overline{F}_{sb}$
m   Stress-Strain curve shape factor.

Allowable Buckling Shear Stress, $\overline{F}_{stb}$

$$\overline{F}_{stb} = \left(\frac{1+v-v^2-v^3}{1-v4}\right)\cdot\overline{F}_{sb} \tag{2.23}$$

where

$$v = 1-\left(\frac{1}{r_o}\right)$$

The associated allowable torque is:

$$\overline{T}_b = \frac{\overline{F}_{stb} \cdot J}{r_o}$$

where

$$J = \frac{\pi\left(r_o^4 - r_i^4\right)}{2}$$

equivalent to F_{sb}, hence iteration is required to obtain a value.

2.8.2 MATERIAL FAILURE UNDER PLASTIC TORSION

The allowable plastic stress at r_o $\overline{F}_{stp} = \frac{3}{4} \cdot \overline{F}_{su} \frac{\left[r_o^3 - r_i^3\right] r_o}{\left[r_o^4 - r_i^4\right]}$ (2.24)

Assuming a constant shear stress, F_{su} through the thickness:

The allowable plastic torque: $\overline{T}_p = \frac{2\pi}{3} \cdot \overline{F}_{su} \left[r_o^3 - r_i^3\right]$ (2.25)

2.9 SAND HEAP ANALOGY FOR TORSIONAL STRENGTH

An experimental method for determining the torsional strength of noncircular sections has been developed by Bendix Aviation and McDonnell Aircraft Company and enables a fairly simple solution to complex shapes as found on aircraft landing components.

The procedure is presented here, as it has been helpful in the past and used by the author for checking complex shapes.

The following is a preamble to the method and is provided for information only.

A solid body, if subject to a severe twisting moment, will deform plastically when the intensity of the shear stress reaches the yield point in shear. The sand heap analogy method is an experimental representation of the stress distribution acting over the cross section of a component in a completely plastic state. The theory predicts that the maximum twisting moment a body can sustain before failure is equal to twice the volume enclosed by the surface, formed by the stress function. The slope of this surface will then be a maximum and therefore analogous to the natural sloping surface formed under gravity by heaping sand on a piece of cardboard having the proportional geometric configuration of the component's cross section.

Mass distribution and shape of the stress curve are important factors in limiting the maximum shear stress for calculation purposes. Experiments have verified the feasibility of using F_{su} as a conservative means of calculating the ultimate twisting moment.

2.9.1 METHOD (SOLID CROSS SECTION)

A horizontally positioned piece of cardboard shaped to the component's cross section is heaped with a fine-grade casting sand, such that a natural roof, defined by gravity, is formed.

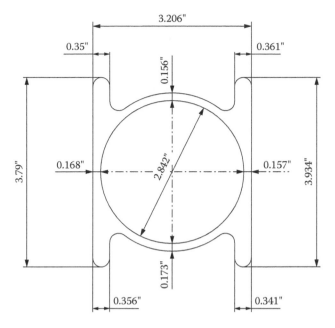

FIGURE 2.13   Sand heap analysis–example section.

In terms of the physical slope and sand volume, the following formula will apply.

$$T = \frac{(2F_{su})2q}{2kC^3} = \frac{q}{(2.54)^3 \, kC^3}(2F_{su}) = \frac{0.061}{kC^3} \, q(2F_{su}) \qquad (2.26)$$

thus

$$T = \frac{0.083 \, q}{C^3}(2\,F_{su})$$

where:

   T = Allowable twisting moment (in.-lb)
   q = Sand volume (cubic centimeters)
   F_{su} = Ultimate shear stress
   C = Scale factor, i.e., double scale, C = 2

Calculation of K and T is as follows:

K value:

Plastic torsion capacity

$$T = \frac{2g \, f_{st}}{254^3 \, KC^3}$$

For a solid circular section:

Plactic torsion capacity

$$T = \frac{4}{3} \frac{f_{st}}{r} J$$

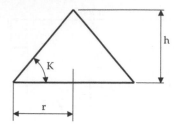

FIGURE 2.14 Angle of repose.

where

$$J = \frac{\pi d^4}{32}$$

$$T = \frac{2\pi f_{st}\, r^3}{3}$$

$\therefore$ For Circle:
$$\frac{2g\, f_{st}}{2.54^3\, KC^3} = \frac{2}{3}\pi\, f_{st}$$

$$K = \frac{3g}{\pi r^3 \cdot 2.54^3 C^3}$$

where:
   g = volume of circle (cm³)
   C = scale factor, or double scale c = 2.0
   K = slope of repose angle, constant for same kind of sand

or:
$$g = \frac{\pi r^2 h (2.54^3 \cdot C^3)}{3} \quad \text{for a circle}$$

$$h = \frac{3g}{\pi r^2 \cdot 2.54^3 \cdot C^3}$$

$$K = \frac{h}{r} = \frac{3g}{\pi r^3 \cdot 2.54^3 \cdot C^3}$$

Calculation of T is as follows:

Measuring the sand volume of the desired section in cm³

$$T = \frac{2g\, f_{st}}{2.54^3\, K\, C^3}$$

where:

f_{st} = ultimate shear allowable (ksi)
g = sand volume of desired section (cm³)
C = scale factor
K = repose angle (can be calculated from (I))

Note that this method requires the shear stress allowable to be measured in ksi (1,000 lbf/in.²).

EXAMPLE 2.1

Consider a cone with a base of 1.50 and a height of 1.063.

$$\text{Total volume of cone} = 42 \text{ cm}^3$$

$$K = \frac{3.42}{\pi(1.5)^3 \cdot 2.54^3 (1)^3}$$

$$= 0.725$$

$$\text{let } h = 0.4 \text{ cm}$$

$$(R = 3.81 \text{ cm} \quad r = 3.231 \text{ cm})$$

$$\text{Volume of frustum} = 11 \text{ cm}^3$$

$$T_{section} = \frac{2 \cdot 11 \cdot 168}{2.54^3 \cdot 0.725 \cdot 1^3}$$

$$= 311.095 \text{ MPa}$$

EXAMPLE 2.2

Figure 2.15 shows a section through a nose landing wheel strut. The original analysis used the sand heap method to establish the allowable torsion to be applied to the section.

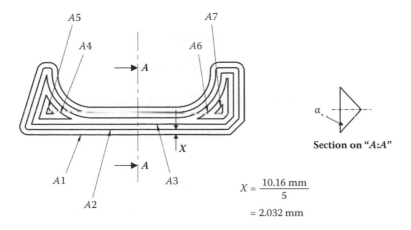

FIGURE 2.15   Section for Example 2.2.

Use a suitable CAD drawing program and setting the outer profile. The profile is then offset at the X dimension. The area is then calculated and tabulated.

By analysis of Figure 2.15, the following individual areas were obtained:

$A_1 = 1{,}317.12$ mm^2

$A_2 = 856.29$ mm^2

$A_3 = 422.87$ mm^2

$A_4 = 45.20$ mm^2

$A_5 = 2.95$ mm^2

$A_6 = 60.19$ mm^2

$A_7 = 4.74$ mm^2

$A_8 = A_4 + A_6 = 47.20$ mm^2 + 60.19 mm^2 = 105.39 mm^2

$A_9 = A_5 + A_8 = 2.95$ mm^2 + 4.74 mm^2 = 7.69 mm^2

$$V_{total} = \left[A_2 + \left(\frac{A_1 - A_2}{2} \right) + A_3 + \left(\frac{A_2 - A_3}{2} \right) + A_8 + \left(\frac{A_3 - A_8}{2} \right) + A_9 + \left(\frac{A_8 - A_9}{2} \right) \right]$$

$$= \left[856.29 + \left(\frac{1317.12 - 856.29}{2} \right) + 422.67 + \left(\frac{856.29 - 422.67}{2} \right) + 105.39 \right.$$

$$\left. + \left(\frac{422.67 - 105.39}{2} \right) + 7.69 + \left(\frac{105.39 - 7.69}{2} \right) \right]$$

$$= 2.032 \tan\alpha \left[2046.98 \right]$$

$$= 4159.47 \tan\alpha$$

Equivalent diameter circle:

$$\frac{1}{3} \pi\, r^3 \tan\alpha = 4159.47 \tan\alpha$$

$$r^3 = \frac{4159.47 \times 3}{\pi}$$

$$r = 15.84 \text{ mm}$$

$$Dia' = 31.67 \text{ mm}$$

$$\text{Torsion allowable} = \frac{p \cdot d^3}{16} \times 1158.32 \text{ MPa}$$

$$= 441.02 \text{ MPa}$$

3 Design and Analysis of Lugs and Shear Pins

3.1 NOTATION

The following notation is used in this design guide.

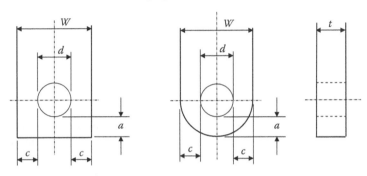

Notation.

The dimensions a, c, d, t, and W are defined in the diagrams above.

σ_{tu} = Ultimate tensile strength of lug material (Pa)
σ_{tuw} = Ultimate tensile strength of lug material with the grain (Pa)
σ_{tux} = Ultimate tensile strength of lug material across grain (Pa)

When the plane of the lug contains both long- and short-transverse grain directions, σ_{tux} is the smaller of the two values.

σ_{ty} = Tensile yield strength of the lug material (Pa)
σ_{tyw} = Tensile yield strength of the lug material with grain (Pa)
σ_{tyx} = Tensile yield strength of the material across grain (Pa)

When the plane of the lug contains both long- and short-transverse grain directions, σ_{tyx} is the smaller of the two values.

σ_{cy} = Compressive yield strength of the bush material (Pa)
P_l = Applied limit load (N)
P_u = Applied ultimate load; this value is usually 1.5 P_l (Pa)
P_y = Applied yield load, this value is usually 1.15 P_l (Pa)
M = Applied bending moment on shear pin (Nm)
α = Angle between direction of loading on lug and the central plane of lug (degree)
P_u' = Allowable ultimate load (N)
P_{bru}' = Allowable ultimate load as determined by shear bearing (N)

P'_{bry} = Allowable yield-bearing load on bushing (N)
P'_{tu} = Allowable ultimate tension load (N)
P'_{tru} = Allowable ultimate transverse load ($\alpha = 90°$) (N)
A_{br} = Projected bearing area (m^2)
A_t = Minimum net section for tension (m^2)
A_{av} = Weighted average area for transverse load (m^2)
K_{br} = Efficiency factor for shear bearing
K_t = Efficiency factor for tension
K_{tru} = Efficiency factor for transverse loading (ultimate)
K_{try} = Efficiency factor for transverse load (yield)
C   = Yield factor
γ = Pin bending moment reduction factor for peaking
r   = $(a - (D/2))/t$
R   = Load ratio in interactive equation
R_{α} = Load ratio for axial load ($\alpha = 0°$)
R_{tr} = Load ratio for transverse load ($\alpha = 90°$)

3.2   INTRODUCTION

Lugs and shear pins are found in a number of aerospace applications, including landing gear attachment points (typical to that shown in Figure 3.1) together with general engineering components, such as the attachment of hydraulic and pneumatic cylinders, etc.

In 1950 an article appeared in the magazine *Product Engineering* titled "Analysis of Lugs and Shear Pins Made from Aluminum or Steel Alloys." The article was written by F.P. Cozzone, M.A. Melcon, and F.M. Hoblit of the Lockheed Aircraft Corporation. There was a follow-up article written in June 1953 in the same magazine entitled "Developments in the Analysis of Lugs and Shear Pins."

These two articles set the ground rules on how lugs and shear pins were to be analyzed and quickly became the standard used in the aerospace industry.

This guide will show the methods for analyzing lugs with either parallel or tapered sides with either round or square end forms, and deals with lug and shear pin failure modes associated with different load conditions covering axial, transverse, oblique, and out-of-plane loading.

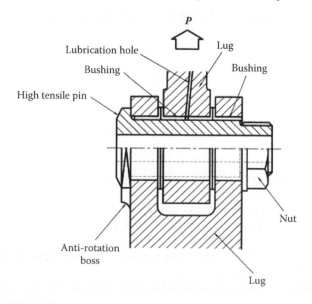

FIGURE 3.1   Typical lug assembly.

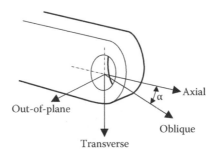

FIGURE 3.2 Lug loading.

This guide also shows how to analyze lugs with clearance fit shear pins and bushes. The stresses associated with the use of interference-fit pins or bushes are also considered.

The shear pin analyses investigate shear pin bending, and shear in double shear joints and pin shear in single shear joints will also be considered.

3.2.1 METHOD

The method described is semiempirical and is applicable to aluminum or steel alloy lugs.

A set of mutually perpendicular axes is applied to the lug, referred to as axial, transverse, and out-of-plane directions. The axial direction is parallel to the lug feature. The transverse direction lies perpendicular to the axial direction in the plane of the lug, and the out-of-plane direction lies perpendicular to the plane of the lug.

The axes intersect at the hole center (see Figure 3.2). The lug is independently analyzed for loads applied in the axial and transverse directions and for the combined oblique loading case. An out-of-plane check is also included.

3.2.2 LOADING

The types of static loading considered in the analysis of lugs and shear pins follow:

Axial
Transverse
Oblique
Out-of-plane

3.2.3 MATERIAL LIMITATIONS

The calculation method may be applied to either steel or aluminum alloy materials.

In the case of axial failure modes it may be applied to lugs manufactured from any metallic material.

3.2.4 GEOMETRIC LIMITATIONS

When the hole diameter (d) is less than the lug thickness (t) it is recommended that the thickness (t) should be set equal to the hole diameter (d) throughout the calculations, except when considering shear pin bending.

When the hole diameter (d) is greater that the lug thickness (8 × t), lug stability should be checked.

The method for determining transverse areas is valid for the lug shapes shown in Figure 3.3.

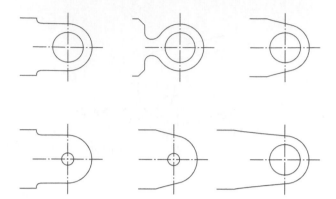

FIGURE 3.3 Types of lugs.

3.2.5 FAILURE MODES

The failure modes (Figure 3.4) follow:

Tension
Shear tear out or bearing (shear bearing)
Yield

3.2.6 NOTES

The margins of safety should be reduced by the relevant fitting or casting factors (see Section 3.4). If both fitting and casting factors are applicable, only the greater of the two need to be considered.

Margins of safety > 0.2 are preferable on lugs for use in aerospace vehicles. In general, in engineering projects where weight is not so important, the margin of safety (MoS) can be increased to 0.3 and above without any detrimental effects.

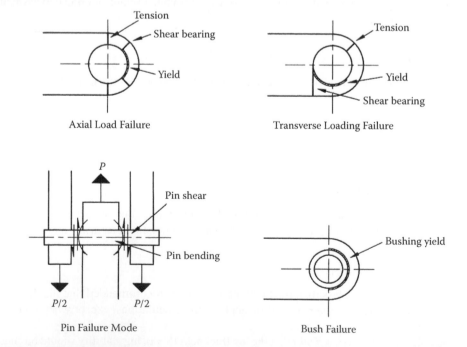

FIGURE 3.4 Lug failure modes.

3.3 ANALYSIS OF LUGS WITH AXIAL LOADING: ALLOWABLE LOADS

A lug-shear pin combination under a tension load can fail in a variety of ways, each of which will require to be evaluated by the methods presented in this section.

Tension across the net section: Here stress concentrations will need to be considered.
Shear tear-out or bearing: These two conditions are closely related and are covered by a single calculation using empirical curves.
Shear acting on the pin: This will be analyzed in the usual way.
Bending of the shear pin: The ultimate strength of the pin is based on the Modulus of Rupture.
Excessive yielding of bushing (if used).
Yielding of the lug is considered to be excessive at a permanent set equal to 0.2 times the shear pin diameter. This condition must always be checked, as it is frequently reached at a lower load than would be anticipated from the ratio of the yield stress (F_{ty}) to the ultimate stress (F_{tu}) for the material.

Note the following:

1. Hoop tension at the tip of the lug is not a critical condition, as the shear-bearing condition precludes a hoop tension failure.
2. The lug should be checked for side loads (due to misalignment, etc.) by conventional beam formulas (see Section 3.4).

3.3.1 ANALYSIS PROCEDURE TO DETERMINE THE ULTIMATE AXIAL LOAD

Calculate e/d, W/d, d/t, $A_{br} = dt$, $A_t = (W - d)/t$ (see Figures 3.5 and 3.6 for nomenclature).
Determine K_{br} using the coordinates e/d and d/t.
The ultimate load for shear-bearing failure, P'_{bru}, is

$$P'_{bru} = K_{br} F_{tu} A_{br} \qquad (3.1)$$

Ultimate load for tension failure:
From Figure 3.6 using W/d, obtain K_t for the lug candidate material.
The ultimate load for tension failure P'_{tu} is

$$P'_{tu} = K_t F_{tu} A_t \qquad (3.2)$$

where F_{tu} = ultimate tensile strength of lug material.

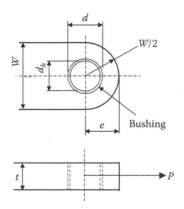

FIGURE 3.5 Detail of lug with bushing.

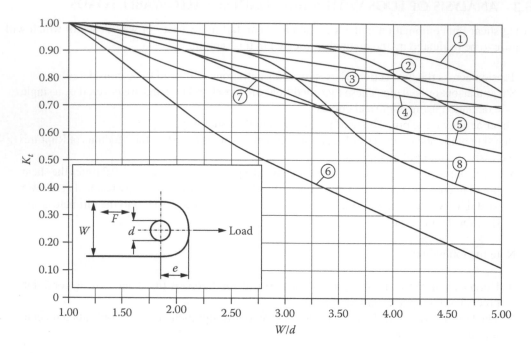

FIGURE 3.6 Effect of grain direction subject to axial load. L, T, N indicates the grain direction "F" in sketch; L = Longitudinal; T = Long transverse; N = Short transverse (normal).

Load for yielding of the lug.

From Figure 3.8 with coordinates of e/d to obtain K_{bry}.

The yield load, P'_y, is

$$P'_y = K_{bry} A_{br} F_{ty} \quad \text{(for } K_{bry} \text{, see Figure 3.8)} \tag{3.3}$$

Load for yielding of the bushing in bearing (if used).

$$P'_{bry} = 1.85 F_{cy} A_{brb} \tag{3.4}$$

where F_{cy} = compressive yield stress of the bushing material.

For the shear pin bending stress see Section 3.7.

Curve 1.

4130, 4140, 4340, and 8630 steel.

2014-T6 and 7075-T6 plate ≤ 0.5 in (12.7 mm) (L, T).

7075-T6 bar and extrusion (L).

2014-T6 hand-forging billet ≤ 144 in² (92903 mm²) (L).

2014-T6 and 7075-T6 die forgings (L).

Curve 2.

2014-T6 and 7075-T6 plate > 0.5 in, ≤ 1.0 in (>12.7 mm, ≤25.4 mm).

7075-T6 extrusion (T, N).

7075-T6 hand-forged billet ≤ 36 in² (23226 mm²) (L).

2014-T6 hand-forged billet > 144 in² (92903 mm²) (L).

2014-T6 hand-forged billet ≤ 36 in² (23226 mm²) (T).

17-4 PH

17-7 PH-THD.

Curve 3.

2024-T6 plate (L, T).

2024-T4 and 2024-T42 extrusion (L, T, N).

Curve 4.

2024-T4 plate (L, T).

2024-T3 plate (L, T).

2014-T6 and 7075-T6 plate > 1.0 in (25.4 mm).

2024-T4 bar (L, T).

7075-T6 hand-forged billet > 36 in² (23226 mm²) (L).

7075-T6 hand-forged billet ≤ 16 in² (10323 mm²) (T).

Curve 5.

195T6, 220T4, and 356T6 aluminium alloy casting.

7075-T6 hand-forged billet > 16 in² (10323 mm²) (T).

2014-T6 hand-forged billet > 36 in² (10323 mm²) (T).

Curve 6.

Aluminium alloy plate, bar, hand forged billet and die forging (N).

Note: for die forgings, N direction exists only at the parting plane.

7075-T6 bar (T).

Curve 7.

18-8 stainless steel, annealed.

Curve 8.

18-8 stainless steel, full hard.

Note: for 1/4, 1/2, and 3/4 hard interpolate between curves 7 and 8.

In Figure 3.7, curve A is a cutoff to be used for all aluminum alloy hand-forged billets when the long-transverse grain direction has the general direction G. Curve B is to be used for all aluminum alloy plate, bar, and hand-forged billets when the short-transverse grain direction has the general direction G, and for die forgings when the lug contains the parting plane in a direction approximately normal to the direction G.

In addition to the limitations provided by curves A and B, in no event shall a K_{br} greater than 2.00 be used for lugs made from 12.7 mm thick or thicker aluminum alloy plate, bar, or hand-forged billets. (See Figure 3.8.)

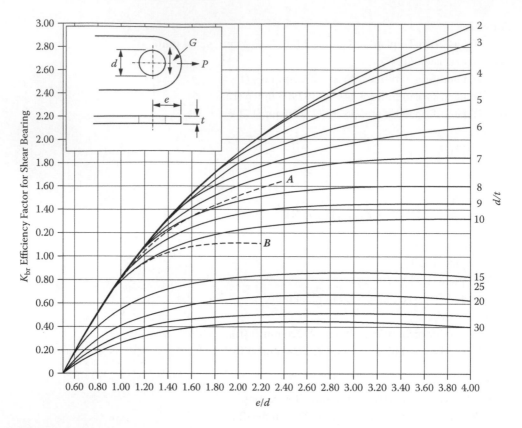

FIGURE 3.7   Shear efficiency factor K_{br}.

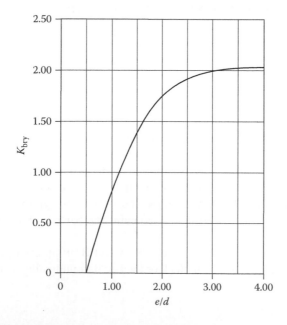

FIGURE 3.8   Factors for calculating yield axial loads attributable to shear bearing of lug and pin.

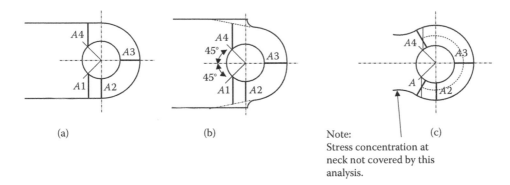

FIGURE 3.9 Transverse areas calculations.

3.4 ANALYSIS OF LUGS WITH TRANSVERSE LOADING: ALLOWABLE LOADS

In the case where the loading being applied to the lug is at $\alpha = 90°$, as in Figure 3.9, the loading is treated as transverse. In order to determine the ultimate and yield loads for the lug, the shape of the lug must be taken into account. This is accomplished by the use of a shape parameter given by

$$\text{Shape parameter} = \frac{A_{av}}{A_{br}}$$

where

A_{br} = the bearing area = dt
A_{vr} = the weighted average area given by:

$$A_{vr} = \frac{6}{\left(\dfrac{3}{A_1}\right)+\left(\dfrac{1}{A_2}\right)+\left(\dfrac{1}{A_3}\right)+\left(\dfrac{1}{A_4}\right)}$$

A_1, A_2, A_3, and A_4 are areas of the lug sections indicated in Figure 3.9.

Obtain the areas A_1, A_2, A_3, and A_4 as follows: A_1, A_2, A_3, and A_4 are measured on the planes indicated in Figure 3.9(a) (perpendicular to the axial center line), except that in a necked lug (Figure 3.9(c)). A_1 and A_4 should be measured perpendicular to the local centerline.

A_3 is the least area on any radial section around the hole. Since the choice of areas and the method of averaging has been substantiated only for the lugs of the shapes shown in Figure 3.10, thought should always be given to ensure that the areas A_1, A_2, A_3, and A_4 adequately reflect the strength of the lug.

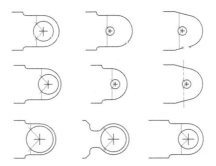

FIGURE 3.10 Substantiated lug shapes.

Obtain the weighted average of these areas using

$$A_{vr} = \frac{6}{\left(\dfrac{3}{A_1}\right) + \left(\dfrac{1}{A_2}\right) + \left(\dfrac{1}{A_3}\right) + \left(\dfrac{1}{A_4}\right)}$$

Compute:

$$A_{br} = dt, \quad \text{and} \quad \frac{A_{av}}{A_{br}}$$

Determine the allowable ultimate load:

$P'_{tru} = K_{tru}A_{br}F_{tux}$

P_{tru} = Allowable ultimate load as determined for the transverse load.

K_{tru} = Efficiency factor for transverse load (ultimate) (see Figure 3.11).

A_{br} = Projected bearing area.

F_{tux} = Ultimate tensile strength of the lug material across the grain.

where:

Yield load P'_y of the lug:

Obtain K_{try} from Figure 3.11.

$$P'_y = K_{try} A_{br} F_{tyx}$$

where:

P'_y = Allowable yield load on the lug
K_{try} = Efficiency factor for transverse load (yield); see Figure 3.11
A_{br} = Projected bearing area
F_{tyx} = Tensile yield stress of lug material across the grain

Determine the allowable yield-bearing load on the bushing using

$P_{bry} = 1.85 F_{cy}A_{brb}$

where:

P_{bry} = Allowable yield-bearing load on bushing
F_{cy} = Compressive yield stress of bushing material
A_{brb} = The smaller of the bearing areas of bushing on pin or bushing on lug (the latter may be smaller as a result of external chamfer on the bushing)

In Figure 3.11 all curves are for K_{tru} except for the one noted as K_{try}.
Note that the curve for 125,000 HT steel agrees closely with test data. Curves for other materials have been obtained by the best available means of correcting for material properties and may possibly be very conservative in some places.

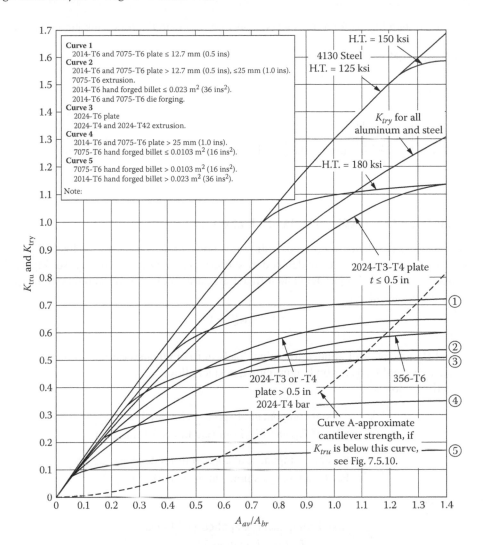

FIGURE 3.11 Efficiency factor for transverse load K_{tru}.

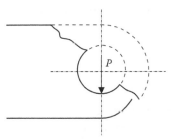

FIGURE 3.12 Lug as a cantilever carrying load.

In no case should the ultimate transverse load be taken as less than that which can be carried by cantilever beam action of the portion of the lug under the load (Figure 3.12). The load that can be carried by cantilever beam action is indicated very approximately by curve A in Figure 3.7; should K_{tru} be below curve A, separate calculations, as a cantilever beam, are warranted.

Investigate pin bending as for axial load (see Section 3.6) with the following modifications:

Take (P'_u) min $= P'_{tru}$.

In the equation:

$$r = \frac{\left[e - \left(\frac{d}{2} \right) \right]}{t} \quad \text{use for the} \quad \left[e - \left(\frac{d}{2} \right) \right] \text{ term the edge distance at a} = 90°$$

3.5   OBLIQUE LOADING: ALLOWABLE LOADING

In the analysis of lugs subject to oblique loading it is convenient to resolve the loading into axial and transverse components; these are denoted by subscripts a and tr, respectively (see Figure 3.13). The two cases are analyzed separately and utilize the results by means of an interaction equation. The interaction equation:

$$R_a^{1.6} + R_{tr}^{1.6} = 1$$

where R_a and R_{tr} are ratios of applied to critical loads in the indicated directions, is to be used for both ultimate and yield loads for both aluminum and steel alloys.
where, for ultimate loads:

$$R_a = \frac{\text{axial component of applied ultimate load}}{\text{smaller of } P'_{bru} \text{ and } P'_{tu} \text{ (from Equations 3.1 and 3.2)}}$$

$$R_{tr} = \frac{\text{transverse component of applied ultimate load}}{P'_{tru} \text{ (from analysis procedure for a} = 90°)}$$

and for yield load:

$$R_a = \frac{\text{axial component of applied yield load}}{P'_y \text{ (fom Equation 3.3)}}$$

$$R_{tr} = \frac{\text{transverse component of applied yield load}}{P'_{try} \text{ (from analysis procedure for a} = 90°)}$$

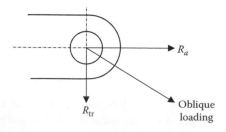

FIGURE 3.13   Oblique loading.

3.5.1 Analysis Procedure

Resolve the applied load into axial and transverse components and obtain the lug ultimate and yield margins of safety from the interaction equation.

$$\text{MoS.} = \frac{1}{\left(R_a^{1.6} + R_{tr}^{1.6} = 1\right)^{0.625}} - 1$$

Check pin shear and bushing yield as in Section 3.7.
Investigate pin bending using the procedure for axial load modified as follows:

$$\left(P_u'\right)_{min} = \frac{P}{\left(R_a^{1.6} + R_{tr}^{1.6} = 1\right)^{0.625}}$$

In the equation

$$\frac{\left[e - (d/2)\right]}{t}$$

use for the [e–(d/2)] term the edge distance at the value of α corresponding to the direction of the load acting on the lug.

3.5.2 Out-of-Plane Loading

In general lugs should not be subject to out-of-plane loading; however, if there is a small component in the R_z direction (Figure 3.14), then it is advisable to check the minimum section for combined bending and shear. This check may be incorporated into the oblique load interaction formula detailed in Section 3.5.1.

The allowable load in the out-of-plane (R_z) direction at section A:A in Figure 3.14 can be expressed as

$$\overline{P}_z = \frac{\left(W - d_h\right) \cdot t \cdot F_{tu}}{\sqrt{\left[\left(\frac{3\,d_h}{K\,t}\right)^2 + \left(\frac{F_{tu}}{F_{su}}\right)^2\right]}}$$

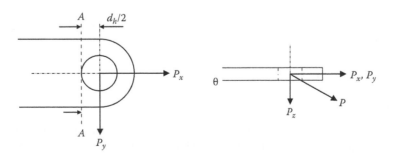

FIGURE 3.14 Out-of-plane loading, load direction, and bending section.

where K is the form factor for bending. The allowable oblique load becomes

$$\bar{P} = \frac{1}{\left[\left(\dfrac{Cos(\theta)Cos(\alpha)}{\bar{P}_{ux\ min}}\right)^{1.6} + \left(\dfrac{Cos(\theta)Sin(\alpha)}{\bar{P}_{uy}}\right)^{1.6} + \left(\dfrac{Sin(\theta)}{P_z}\right)^{1.6}\right]^{0.625}}$$

Estimate the oblique angle θ. Generally, if no design data are available, the oblique angle may be assumed to be at least 10° to allow for any misalignment, etc.

3.6   BEARING AT LUG-TO-PIN OR -BUSH INTERFACE

The allowable bearing loads at the lug-to-pin interface are given below. A combined shear tear-out and bearing check has previously been performed for axial, transverse, and oblique loadings, respectively. However, this check may be a separate requirement, particularly for lugs subject to compression.

MIL HDBK 5 provides values of allowable ultimate and yield-bearing stress for (a/d) values of 1.5 and 2.0, valid for values of (d/t) up to 5.5. Bearing allowables for other values of (a/d) can be estimated by linear interpolation, as shown below.

$\dfrac{a}{d}$ range	Equation for bearing strength
$0.5 < \dfrac{a}{d} \leq 1.5$	$\bar{F}_{br_n} = \left(\dfrac{a}{d} - 0.5\right)\bar{F}_{br_{1.5}}$
$1.5 < \dfrac{a}{d} < 2.0$	$\bar{F}_{br_n} = \bar{F}_{br_{1.5}} + 2\left(\dfrac{a}{d} - 1.5\right)\left(\bar{F}_{br_2} - \bar{F}_{br_{1.5}}\right)$
$\dfrac{a}{d} \geq 2.0$	$\bar{F}_{br_n} = \bar{F}_{br_2}$

Ultimate bearing:

$$\bar{P}_{bru} = \bar{F}_{bru} \cdot d_{pin} \cdot t$$

Yield bearing:

$$\bar{P}_{bry} = \bar{F}_{bry} \cdot d_{pin} \cdot t$$

Note that the dimension a, the distance from the center of the hole to the edge of the lug, is denoted as e in MIL HDBK 5.

3.7   SHEAR PIN ANALYSIS

3.7.1   SHEAR PIN BENDING IN DOUBLE SHEAR JOINT

In static tests of a single bolt fitting, failure of the shear pin due to bending failure will not be shown to be a factor in the failure of the lug. However, it is important to provide sufficient bending strength to ensure permanent bending deformation does not occur when subject to the limit loads so that the shear pin can be readily removed for inspection and maintenance operations.

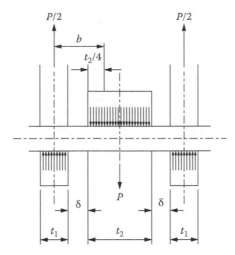

FIGURE 3.15 Pin moment arm.

Weakness in the shear pin can cause peaking up non-uniform-bearing loads on the lugs, influencing the lug tension and shear strength. The big unknown factor in shear pin bending is the true value of the bending moment acting on the pin because the moment arm to the resultant bearing forces is difficult to quantify.

An approximate method that is commonly used to determine the moment arm on the shear pin in a double shear joint is shown below:

$$\text{Moment arm (b)} = 0.5t_1 + 0.25t_2 + d$$

If there is no gap ($\delta = 0$) between the lug faces (see Figure 3.15) the moment arm is taken as

$$b = 0.5\,t_1 + 0.25\,t_2$$

The applied shear pin bending moment is given by

$$m = b \times \frac{P}{2}$$

where P is the minimum failure load for the center lug or twice that of the outer lugs, and b is the moment arm (see Figure 3.15). It is assumed that the maximum shear pin load is the minimum lug failure load.

Calculate bending stress from

$$\sigma_m = \frac{M}{Z}$$

where

$$Z = \frac{\pi}{32}\,d^3$$

M = Bending moment

If the pin strength is found to be inadequate by the above check, then use may be made of the tendency of the pin loads to peak near the shear faces, thereby reducing the bending moment arm.

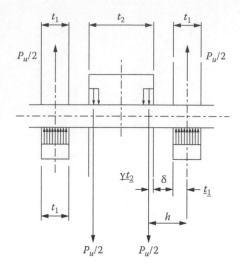

FIGURE 3.16   Pin moment arm with clearance.

No peaking in the outer lugs can occur unless a substantial head or nut bears firmly on the outer lugs. The reduction in the moment arm will apply only to the central lug.

A reduction in the shear pin bending up to 25% can be expected.

3.7.1.1   Shear Pin Bending: Load Peaking between Center Lug and Pin

If it is desired to take into account the reduction of pin bending that results from load peaking, as shown in Figure 3.16, the moment arm is obtained as follows: Calculate the inner lug r (use d_b instead of d, if bushing is used).

$$r = \frac{\left[\left(\frac{a}{D}\right) - \frac{1}{2}\right]D}{t_2}$$

Take the smaller of P_{bru} and P_{tu} for the inner lug as $(P_u)_{min}$ and calculate

$$\frac{(P_u)_{min}}{A_{br}F_{tux}}$$

From Figure 3.17 using $\dfrac{(P_u)_{min}}{A_{br}F_{tux}}$ and r obtain the reduction factor γ for peaking and calculate the moment arm.

The maximum bending moment in the shear pin:

$$h = \left(\frac{t_1}{2}\right) + \delta + \gamma\left(\frac{t_2}{4}\right)$$

and calculate the bending stress in the shear pin that results from M, assuming an MD/2I distribution (where I = moment of Inertia of the pin) and its Margin of Safety (MoS) as:

$$MoS = \frac{\left(\dfrac{MD}{2I}\right)}{F_b} - 1$$

where F_b = the Modulus of Rupture determined from Figure 3.18 or other sources.

$$M = \frac{P_u h}{2}$$

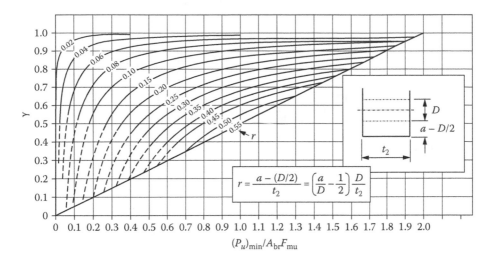

FIGURE 3.17 Peaking factors for pin bending. (Dashed lines indicate where the theoretical curves are not substantiated by test data.)

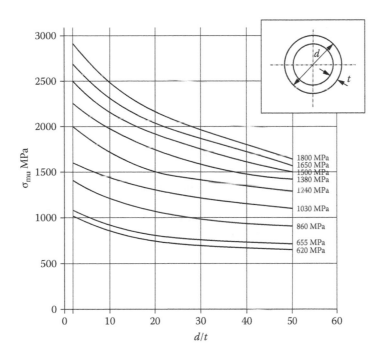

FIGURE 3.18 Bending Modulus of Rupture.

3.7.1.2 Shear Pin Bending, Including Excess Strength of Lug

This approach should only be used if there is an excess margin in the strength of the lug and the pin is still inadequate.

In this circumstance it may be assumed that a portion of the lug thickness in the joint is inactive (see Figure 3.19). The active thicknesses are chosen by trial and error to give approximately equal margins of safety (MoS) for the lug and the pin.

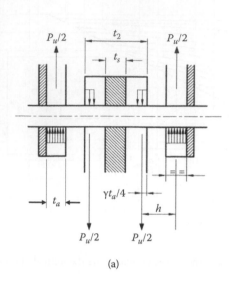

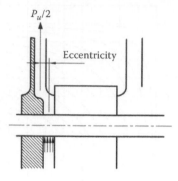

t_s = Shaded inactive thickness

t_a = Active thickness = $t_2 - t_s$

(a)   (b)

FIGURE 3.19   Load peaking.

Using the smaller of

$$\frac{2c}{d} \frac{\lambda_t}{R_{tu}}$$

and

$$\frac{2a}{d} \frac{\lambda_s}{R_{tu}}$$

for the inner lug, obtain the reduction factor γ for peaking where:

$$r = \frac{a - d/2}{t_a}$$

and t_a is the active thickness shown in Figure 3.19(a), noting that γ applies only to the inner lug. Calculate the moment arm h as indicated below:

Maximum shear pin bending moment $M = \dfrac{P_u h}{2}$ as before.

If the eccentricity occurs in an outer lug (see Figure 3.19b), the structure must be capable of withstanding the induced bending; otherwise, the shear pin must be strong enough for the analysis by the method given in Section 3.7.1.1.

3.7.2   PIN SHEAR

The allowable shear load for a solid shear pin in a double shear joint:

$$P_S = K_s \overline{F_s} A_{pin}$$

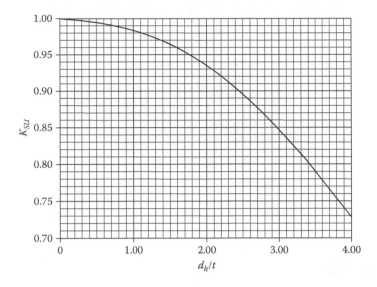

FIGURE 3.20   Allowable shear stress factor for pin in double shear.

where

$$A_{pin} = \frac{\pi \times d_{pin}^{\,2}}{4}$$

In double shear, the applied load will be the load in the outer lugs or half the load in the center lug.

The allowable shear stress $\overline{F}_s$ is to be factored by K_s to allow for the variation in the ratio (d_h/t). K_{su} can be determined from Figure 3.20. Limitations on the use of the curve are as follows:

$$\frac{d_h}{t} < 4.0 \qquad t_o > 0.75\, t_i$$

For a single shear joint the above equation will still apply, but in this case Ksu = 1.0.

3.8   BUSH ANALYSIS

The allowable yield load on bushing:

$$P_{bry} = 1.85\,\overline{F}_{cy}\,A_{brb}$$

where A_{brb} is the smaller bearing area of the busing on the shear pin or bushing on lug.

The bearing area on the lug may be less if the bushing has been chamfered. $\overline{F}_{cy}$ is the allowable compressive yield stress on the bush material.

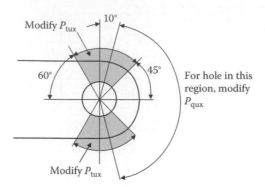

FIGURE 3.21   Modification of allowable load for the presence of a lubrication hole.

3.9   SPECIAL CASES

3.9.1   OIL HOLES

3.9.1.1   Axial Load

The calculation of P_{tux} or P_{qux}, respectively, should be modified depending on the location of the lubrication hole (see Figure 3.21). For the modification of these parameters the lug tension and bearing areas should be based on the lug thickness minus the lubrication hole diameter. The value of K_{qux} will need to be recomputed to account for the reduction in thickness.

3.9.1.2   Transverse Load

Obtain P_{uy} neglecting the lubrication hole and multiply by:

$$0.90\left(1 - \frac{d_{\text{Lubrication hole}}}{t}\right)$$

3.9.1.3   Oblique Load

In this case proceed as normal and obtain T_{tux} and P_{qux} as per Section 3.9.1.1 and obtain P_{uy} as per Section 3.9.1.2. Disregard the lubrication hole in calculating the yield margin.

3.9.2   ECCENTRIC HOLE

If the hole is laterally located (see Figure 3.22(a)), i.e., e_1 is less than e_2, the ultimate and yield lug loads are determined by calculating P'_{bru}, P'_{tu}, and P'_y for the equivalent lug shown in Figure 3.22(b) and multiplying by the factor below:

$$\text{Factor} = \frac{e_1 + e_2 + 2d}{2e_2 + 2d}$$

3.9.3   MULTIPLE SHEAR CONNECTION

A shear pin-lug combination having a geometry shown in Figure 3.23 can be solved according to the following procedure:

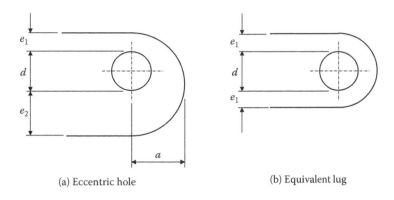

(a) Eccentric hole   (b) Equivalent lug

FIGURE 3.22 Equivalent lug for a lug with an eccentric hole

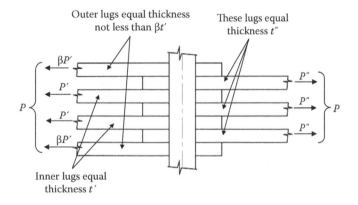

FIGURE 3.23 Multiple shear connections.

TABLE 3.1

Pin Shear and Moment Arms in Multiple Shear Lugs

Number of Lugs Including Side Lugs	β	Pin Shear	Moment Arm
5	0.35	0.50 P'	0.28 (t' + t'')/2
7	0.40	0.53 P'	0.33 (t' + t'')/2
9	0.43	0.54 P'	0.37 (t' + t'')/2
11	0.44	0.54 P'	0.39 (t' + t'')/2
Infinity	0.50	0.50 P'	0.50 (t' + t'')/2

---

1. The load carried by each individual lug can be determined by distributing the total applied load among the lugs, as shown in Figure 3.23. β is obtained from Table 3.1.
2. The maximum shear load acting on the shear pin is also given in Table 3.1.
3. The maximum bending moment on the shear pin is given by the formula

$$M = \frac{P \times \beta}{2}$$

where β is given in Table 3.1.

3.10   STRESSES DUE TO INTERFERENCE-FIT PINS AND BUSHES

The insertion of circular interference-fit shear pins or bushes of similar or dissimilar materials in flat bars, strips, or lugs will introduce elastic stresses in these components. The analysis will be treated in two parts: solid shear pins (Section 3.10.1) and bushes (Section 3.10.2).

The method is based upon a two-dimensional analysis and will only give an average stress through the thickness of the lug. Local variations away from the average values will most likely be nonsignificant unless taper pins are used.

It is assumed that the material will be isotropic (obeys Hooke's law), but the method can be used for most engineering materials without any significant error.

3.10.1   Solid Circular Interference-Fit Shear Pins

In the case of an interference-fit shear pin fitted in an infinitely wide plate, i.e., $d_h/W = 0$, the circumferential tensile, radial compressive, and shear stresses will be uniform around the hole.

For a plate of finite width, the stresses will vary around the hole, with the maximum value being dependent upon d_h/W.

Referring to Figure 3.24, at the points marked B the tensile stress σ, shear stress q, and interference pressure p will approach the values as determined for a shear pin in an infinite plate. The maximum tensile and shear stresses will occur at the points marked A, although the shear stress will only be slightly above its value as predicted at points B. The interference pressure will be at a maximum at points B and a minimum at points A.

> At point A:
>
> Maximum tensile and shear stresses
> Minimum pressure
>
> At point B:
>
> Tensile and shear stresses will tend to reach values as predicted for an infinite plate
> Maximum pressure

An estimate of the interface pressure at B is given by

$$\frac{p}{E_1 e} = \left[1 - \left(\frac{d_h}{W}\right)^2\right]\left(\frac{q}{E_1 \times e}\right)$$

Shear stress is given by:

$$\frac{q}{E_1 e} = \frac{1}{2 + \left[\dfrac{E_1\left(1 - \mu_p\right)}{E_p} - \left(1 - \mu_1\right)\right]\left[1 - \left(\dfrac{d_h}{W}\right)^2\right]}$$

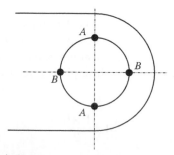

FIGURE 3.24   Points of high stress in bore.

Tensile stress is given by:

$$\frac{\sigma}{E_1 \cdot e} = \frac{q}{E_1 \cdot e}\left[1+\left(\frac{d_h}{W}\right)^2\right]$$

These formulas are derived from a pin in a ring, but they may be used with reasonable accuracy for a pin in a lug provided $(d_h/W) < 0.8$.

3.10.2 INTERFERENCE-FIT BUSHES

In the following analysis d_b is the bush internal diameter and K is the bush shape factor.

The maximum tensile stress occurs at the points marked A in Figure 3.24. The maximum shear stress is given by

$$q_{max} = \frac{(\sigma + p)}{2}$$

The form of the end of a bushed lug affects the interference pressure and maximum stress. Figure 3.25 provides a correction factor to (d_h/W) for the effect of round- and square-ended lugs.

The interference pressure is given by

$$\frac{p}{E_1\, e} = \frac{1}{\left[\dfrac{1+\left(K\, d_h/W\right)^2}{1-\left(K\, d_h/W\right)^2}\right]+\mu_1+\dfrac{E_1}{E_b}\left[\dfrac{1+\left(d_b/d_h\right)^2}{1-\left(d_b/d_h\right)^2}-\mu_b\right]}$$

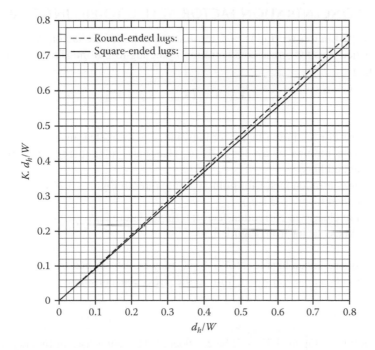

FIGURE 3.25 Bush shape factor.

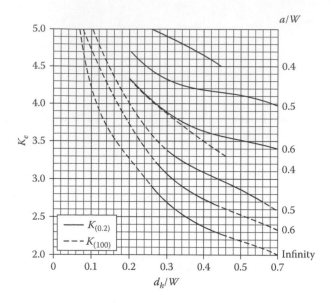

FIGURE 3.26   Stress concentration factor for round lugs.

and tensile stress is given by:

$$\frac{\sigma}{E_l e} = \frac{p}{E_l e}\left[\frac{1+\left(d_b/d_h\right)^2}{1-\left(d_b/d_h\right)^2}\right]$$

3.11   STRESS CONCENTRATION FACTOR AT LUG-TO-PIN INTERFACE

The method presented in this section is to determine the axial failure factor, K_{tux}.

For axially loaded lugs the stress concentration at the lug-to-pin interface is defined by the following equation:

$$K_e = \left(K_{0.2} + \eta_e\left(K_{100} - K_{0.2}\right)\right)K_{th}$$

where:

K_e = Hole stress concentration factor for lugs with e percent clearance
$K_{0.2}$ = Stress concentration factor for 0.2% d_h clearance; refer to Figure 3.26 for round and square-ended lugs
K_{100} = Stress concentration factor for point loading; refer to Figures 3.26 and 3.27 for round and square-ended lugs (e ≤ 0.1 d_h)
η_e = Lug-to-pin clearance correction factor; refer to Figure 3.29
K_{th} = Lug thickness factor from Figure 3.30

In the case of square-ended lugs (e < 0.1% d_h) K_e may be determined from Figure 3.28.
This case represents a practical lower limit to the stress concentration.

3.12   EXAMPLES

In this example, a fitting (Figure 3.31) will be considered. This comprises of two lugs that connect to two struts. The fitting is used in an engine nacelle and is subject to engine heat.

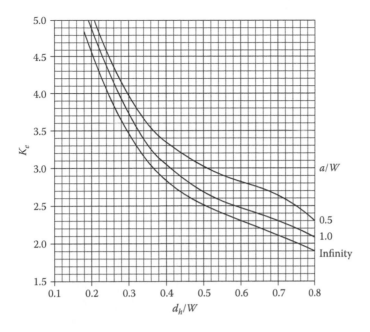

FIGURE 3.27 $K_{0.2}$ stress concentration factor for square lugs.

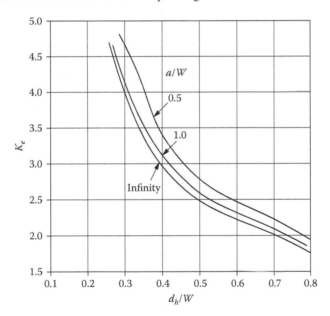

FIGURE 3.28 $K_{(e<0.1)}$ stress concentration factor for square lugs.

Considering the dimensional and material properties of the lug:

Principal Lug Dimensions
    a   = 30.0 mm
    c   = 22.00 mm $(a - d_h/2)$
    d_h = 17.50 mm (hole diameter with bush) (included for repairability)
    d   = 16.00 mm (hole diameter without a bush)
    W = 60.00 mm

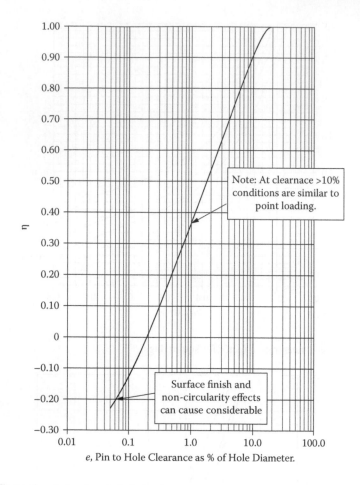

FIGURE 3.29   Correction factor for pin to hole clearance.

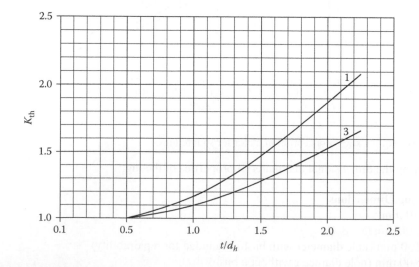

FIGURE 3.30   Effect of lug thickness.

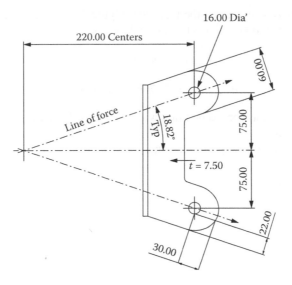

FIGURE 3.31 Detail of fitting.

R = 30.00 mm (W/2)
t = 7.50 mm
r = 8.75 mm (d_h/2)

Material properties and operating temperatures:

Material: 15-5PH Stainless Steel Bar/Forging (AMS 5659) Heat Treated to H1025

The in-flight operating temperature in the engine nacelle	T_{op}	= 150°C
Ultimate tensile stress with thermal factor for 150°C	F_{tu}	= 972.51 MPa
Yield tensile stress with thermal factor for 150°C	F_{ty}	= 909.42 MPa
Yield compressive stress with thermal factor for 150°C	F_{cy}	= 877.50 MPa
Ultimate shear stress with thermal factor for 150°C	F_{su}	= 608.60 MPa
Ultimate bearing stress with thermal factor for 150°C	F_{bru}	= 1,650.12 MPa
Yield-bearing stress with thermal factor for 150°C	F_{bry}	= 1,323.86 MPa
Young's modulus at 150°C	E	= 192.57 GPa

The table above summarizes the material properties used for the construction of the lug, and the allowables cover both longitudinal and transverse directions. Therefore, the analysis does not have to consider grain directions.

Note that the shear pin is made up of a nut-and-bolt assembly.

Summary of Allowables' Nut and Bolt Strengths

Bolt maximum ultimate shear	32.47 N
Bolt maximum ultimate tension	45.37 N
Nut maximum ultimate tension	32.87 N

The bolt is manufactured from A286, and at the operating temperature of 150°C, there is no temperature knock-down factor to be applied.

Consider the axial tension acting on the lug. From Figure 3.31 it is clearly seen that the load acting on the lug is considered to be axial. The nominal load is

$$P_{tensile} = 76.33\text{kN tensile}$$

and

$$P_{compresssive} = 76.70\text{kN compressive}$$

This analysis will consider the bore to be fitted with a bush, as this will give the minimum strength in the ligaments of the lug.

The fitting factor (ff) used throughout this analysis is 1.15.

From Figure 3.34,

a $= 30.00$ mm
$d_h = 17.50$ mm
t $= 7.50$ mm

Hence:

$a/d_h = 1.875$
$d_h/t = 2.133$

The allowable axial load:

$$P_{tu} = K_t \times F_{tu} \times A_t$$

$$P_{tu} = 0.9231 \times 972.51\,\text{MPa} \times 318.75\,\text{mm}^2$$

$$P_{tu} = 285498.5\,\text{N}$$

$$RF = \frac{P_t}{F_a \times ff}$$

$$= \frac{285498.5\,\text{N}}{76330\,\text{N} \times 1.15}$$

$$RF = 3.25$$

Consider the allowable shear-bearing load.
K_{br} is first deduced from Figure 3.7.

$$e/d = 1.87$$

$$d/t = 2.133$$

$$P_{qu} = K_{br} \times A_{br} \times F_{tu}$$

$$= 1.824 \times 131.25\,\text{mm}^2 \times 972.51\,\text{MPa}$$

$$P_{qu} = 232818.89\,\text{N}$$

$$RF = \frac{P_{qu}}{F_a \times ff}$$

$$= \frac{232.819\,\text{N}}{76330\,\text{N} \times 1.15}$$

$$RF = 2.65$$

Consider the yield/proof of the lug. This section checks the minimum proof strength for either tension or shear bearing. (The critical bore size used is d_h, with bush included).

where:
P_{px} = Allowable yield load in the axial direction
K_{bry} = Axial yield factor
F_{ty} = Minimum allowable tensile yield stress, with respect to the grain direction in plane of the lug
F_{tu} = Minimum allowable ultimate tensile stress, with respect to the grain direction in plane of the lug

$$P_{px} = K_{bry} \times \frac{F_{ty}}{F_{tu}} \times \left(P_{tu}, P_{qu}\right) \quad \text{(This is checked for the larger of } P_{tu} \text{ and } P_{qu})$$

$$K_{bry} = 1.586 \text{ (from Figure 3.8)}$$

$$F_{tu} = 972.51 \text{ MPa}$$

$$F_{ty} = 909.42 \text{ MPa}$$

$$P_{tu} = 285498.5 \text{ N}$$

$$P_{qu} = 232818.89 \text{ N}$$

$$P_{px} = 1.586 \times \frac{909.42 \text{ MPa}}{972.51 \text{ MPa}} \times 285498.5 \text{ N} \quad (P_{tu} \text{ is larger than } P_{qu})$$

$$= 423425.92 \text{ N}$$

Safety Factor (SF = 1.5 (proof)

$$RF = \frac{P_{px} \times SF}{F_a \times ff}$$

$$= \frac{423425.92 \text{ N} \times 1.5}{76330 \text{ N} \times 1.15}$$

$$RF = 4.824$$

FURTHER READING

Cozzone, F.P., Mclcon, M.A., and Hoblit, F.M. Analysis of lugs and shear pins made from aluminum or steel alloys. *Product Engineering* 1950.

Melcon, M.A., and Hoblit, F.M. Developments in the analysis of lugs and shear pins. *Product Engineering* 1953.

Niu, Michael C. *Airframe structural design*. Hong Kong: Conmilit Press Ltd., 1999.

4 Mechanical Fasteners

4.1   THREADED FASTENERS

This chapter covers threaded fasteners. These are a very flexible method for attaching components as they allow future easy disassembly. Threaded fasteners are designed principally for tensile loads but will accept a small level of shear load. This chapter begins with a general description of threads and then progresses to cover pre-tension diagrams, and so forth.

4.2   BASIC TYPES OF THREADED FASTENERS

Bolt

A bolt is a threaded fastener that passes through clearance holes in mating members and is secured with a nut at the end opposite the head of the bolt (Figure 4.1).

Screw

A screw is a threaded fastener that passes through a clearance hole in one member and into a threaded hole in the other mating member (Figure 4.2).

Threaded fasteners are available in a wide variety of head shapes (Figure 4.3), including some of the following.

4.3   THREAD STANDARDS

There are several national and international standards covering fastener threads. These include those listed in Table 4.1.

4.4   THREAD PROFILES

The pitch line or diameter is located at 0.5 the height of the theoretical sharp v-thread profile (Figure 4.4).

4.5   THREAD SERIES

The number of threads per unit length distinguishes groups of diameter-pitch combinations from each other when applied to a specific diameter.

ISO Metric Series. Metric system of diameters, pitches, and tolerance/allowances. This system is now the most common thread system in Europe and many parts of the world. There are two series of metric thread: metric coarse and metric fine.

Unified Coarse Thread Series (UNC). Most commonly used in the majority of bolts, screws, and nuts in general engineering applications. This system is most commonly used in the United States and in Europe on old designs not yet converted to metric systems.

Unified Fine Thread Series (UNF). This system is used when there is a requirement for a finer pitch of thread, i.e., when the length of thread engagement is restricted.

Unified Fine J Series (UNJ). This series offer a finer pitch than that of the UNF series. Its most common application is in the aerospace industry.

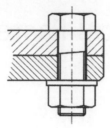

FIGURE 4.1   Bolt.

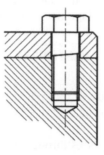

FIGURE 4.2   Screw.

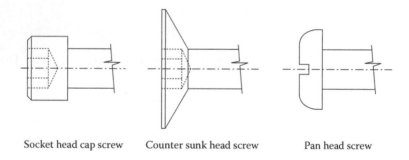

Socket head cap screw   Counter sunk head screw   Pan head screw

FIGURE 4.3   Types of heads.

TABLE 4.1
National and International Standards

Country	Standard Number	Description
UK	BS 3643	ISO metric screw threads
UK	BS 1500	Unified thread series
U.S.	B1.1-1949	Unified threads
U.S.	ANSI B1.1-1989/ASME B1.1-1989	Revision to B1.1-1949
U.S.	ANSI B1.13M-1983 (R1989)	Metric threads (M series)

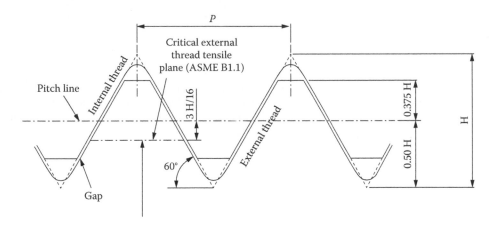

FIGURE 4.4 Typical thread profile.

4.6 THREAD DESIGNATIONS

4.6.1 METRIC SERIES

The following is an example of how a metric thread combination is described in a drawing or a specification.

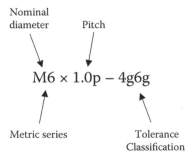

4.6.2 IMPERIAL (INCH SERIES)

The following is how an imperial thread combination is called up on a drawing or a specification.

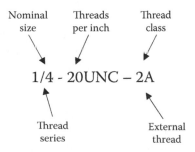

TABLE 4.2
ISO Metric Course Thread

Size Designation	Nominal Major Diameter mm	Pitch Diameter mm	Effective Diameter mm	Minor Diameter mm	Minor Diameter Area mm²	Depth mm	Shear Area External mm²	Shear Area Internal mm²
M3	3.00	0.50	2.675	2.387	4.473	0.307		
M3.5	3.50	0.60	3.110	2.764	6.000	0.368	4.674	6.606
M4	4.00	0.70	3.545	3.141	7.750	0.429	5.473	7.768
M5	5.00	0.80	4.480	4.018	12.683	0.491	7.073	9.997
M6	6.00	1.00	5.350	4.773	17.894	0.613	8.645	12.189
M8	8.00	1.25	7.188	6.466	32.841	0.767	12.160	16.826
M10	10.00	1.50	9.026	8.160	52.292	0.920	15.578	21.473
M12	12.00	1.75	10.863	9.853	76.247	1.074	18.975	26.113
M14	14.00	2.00	12.701	11.546	104.710	1.227	22.422	31.028
M16	16.00	2.00	14.701	13.546	144.120	1.227	26.095	35.563
M18	18.00	2.50	16.376	14.933	175.135	1.534	29.525	40.754
M20	20.00	2.50	18.376	16.933	225.189	1.534	33.276	45.379
M22	22.00	2.50	20.376	18.933	281.527	1.534	37.027	50.004
M24	24.00	3.00	22.051	20.319	324.274	1.840	40.458	54.999
M27	27.00	3.00	25.051	23.319	427.095	1.840	46.169	62.997
M30	30.00	3.50	27.727	25.706	518.990	2.147	51.633	69.537
M33	33.00	3.50	30.727	28.706	647.195	2.147	57.419	76.603
M36	36.00	4.00	33.402	31.092	759.274	2.454	63.091	84.043

Note: All dimensions in mm.

TABLE 4.3
UNC Threads

Designation	O Diameter	Core Diameter Nut	Core Diameter Bolt	Pitch	Depth	Effective Diameter
1/4—20 UNC	0.2500	0.1959	0.1887	0.05000	0.03067	0.2175
5/16—18 UNC	0.3125	0.2524	0.2443	0.05556	0.03408	0.2764
3/8—16 UNC	0.3750	0.3073	0.2983	0.06250	0.03834	0.3344
7/16—14 UNC	0.4375	0.3602	0.3499	0.07143	0.04382	0.3911
1/2—13 UNC	0.5000	0.4167	0.4056	0.07692	0.04719	0.4500
9/16—12 UNC	0.5625	0.4723	0.4603	0.08333	0.05112	0.5084
5/8—11 UNC	0.6250	0.5266	0.5135	0.09091	0.05577	0.5660
3/4—10 UNC	0.7500	0.6417	0.6273	0.10000	0.06134	0.6850
7/8—9 UNC	0.8750	0.7547	0.7387	0.11111	0.06816	0.8028
1.0—8 UNC	1.0000	0.8647	0.8466	0.12500	0.07668	0.9188
1 1/8—7 UNC	1.1250	0.9704	0.9497	0.14286	0.08763	1.0322
1 1/4—7 UNC	1.2500	1.0954	1.0747	0.14286	0.08763	1.1572
1 3/8—6 UNC	1.3750	1.1946	1.1705	0.16667	0.10224	1.2667
1 1/2—6 UNC	1.5000	1.3196	1.2955	0.16667	0.10224	1.3917
1 3/4—5 UNC	1.7500	1.5335	1.5046	0.20000	0.12269	1.6201
2—4 1/2 UNC	2.0000	1.7594	1.7274	0.22222	0.13632	1.8557

Note: All dimensions in inches.

4.7 MATERIAL AND STRENGTH DESIGNATIONS

TABLE 4.4
Metric Grades of Steels for Fasteners

Grade	Fastener Size	Tensile Strength MPa	Yield Strength MPa	Proof Strength MPa
4.6	M5 to M36	400	240	225
4.8	M1.6 to M16	420	340[a]	310
5.8	M5 to M24	520	415[a]	380
8.8	M16 to M36	830	660	600
9.8	M1.6 to M16	900	720[a]	650
10.9	M6 to M36	1,040	940	830
12.9	M1.6 to M36	1,220	1,100	970

[a] Yield strength is approximate and is not included in the standard.

TABLE 4.5
SAE Grades of Steels for Fasteners

Grade Number	Fastener Size (in)	Tensile Strength (ksi)	Yield Strength (ksi)	Proof Strength (ksi)	Head Marking
1	1/4 to 11/2	60	36	33	None
2	1/4 to 3/4	74	57	55	None
	>3/4 to 11/2	60	36	33	
4	1/4 to 11/2	115	100	65	None
5	1/4 to 1.0	120	92	85	3 points
	>1.0 to 11/2	105	81	74	
7	1/4 to 11/2	133	115	105	5 points
8	1/4 to 11/2	150	130	120	6 points

TABLE 4.6
ASTM Standard for Fastener Steels

ASTM Grade	Fastener Size in.	Tensile Strength (ksi)	Yield Strength (ksi)	Proof Strength (ksi)	Head Marking
A307	1/4 to 4	60	—	—	None
A325	1/2 to 1	120	92	85	A325
	>1.0 to 1 1/2	105	81	74	
A354-BC	1/4 to 2 1/2	125	109	105	BC
A354-BD	1/4 to 2 1/2	150	130	120	6 points
A449	1/4 to 1.0	120	92	85	
	> 1.0 to 1 1/2	105	81	74	
	>1 1/2 to 3.0	90	58	55	
A574	0.060 to 1/2	180	—	140	(Socket head
	5/8 to 4.0	170	—	135	cap screws)

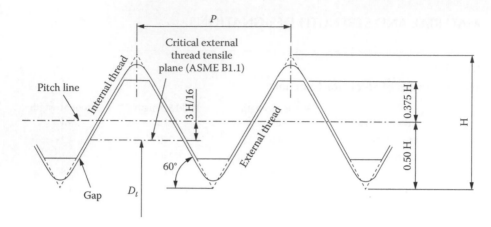

FIGURE 4.5 Tensile stress area.

4.8 TENSILE AND SHEAR STRESS AREAS

4.8.1 TENSILE STRESS AREA

The average axial stress in a fastener is calculated using a tensile stress area (Figure 4.5).

$$f_{ave} = \frac{F}{A_t}$$

$$A_t = \frac{\pi}{4}\left(\frac{D_r + D_p}{2}\right)^2$$

where
  f_{ave} = Average axial stress
  F   = Axial force
  D_r = Root diameter
  D_p = Pitch diameter
  A_t = Tensile stress area

An unthreaded rod having a diameter equal to the mean of the pitch diameter and minor diameter will have the same tensile strength as the threaded rod.

$$D_t = d_b - 2\left(\frac{3}{8}H + \frac{3}{16}H\right)$$

$$H = \frac{1}{2n}\cdot\tan(60°)$$

$$D_t = d_b - \frac{\tan(60°)}{n}\left(\frac{3}{8} + \frac{3}{16}\right)$$

$$D_t = d_b - \frac{9\sqrt{3}}{16n}$$

$$A_t = \frac{\pi}{4}D_t^2$$

$$A_t = \frac{\pi}{4}\left(d_b - \frac{0.9743}{n}\right)^2$$

where
  D_t = Diameter at critical plane
  d_b = Diameter of fastener
  H = Theoretical height of thread
  n = 1/p (threads per unit length)

This is the formula used by manufactures of imperial fasteners to publish the tensile area in their catalogs.

$$A_t = \frac{\pi}{4}\left(d_b - \frac{0.9328}{n}\right)^2$$

This formula is used to obtain a similar result for metric threads.

4.8.2 SHEAR AREA OF EXTERNAL THREADS

The interaction between mating threads needs to be considered to establish the shear area of an external thread (Figure 4.6).

$$A_{s,c} = \pi \cdot K_{n,max} \cdot t_e \cdot n$$

$$\tan(30°) = \frac{0.5\, t_e}{0.75\, H - gap}$$

$$H = \frac{1}{2n}\tan(60°) = \frac{\sqrt{3}}{2n}$$

$$t_e = 2\cdot\tan(30°)\left(0.75\frac{\sqrt{3}}{2n} - gap\right)$$

$$gap = \frac{1}{2}\left(K_{n,max} + \frac{1}{2}\frac{\sqrt{3}}{2n} - E_{s,min}\right)$$

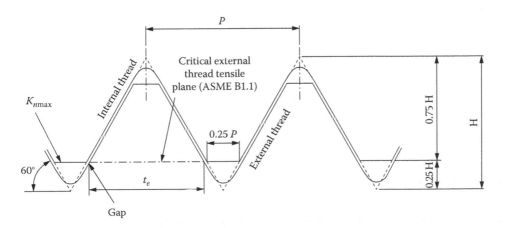

FIGURE 4.6 Shear area of external thread.

where:

$A_{s,c}$ = Shear area of external thread
$K_{n,max}$ = Maximum minor diameter of internal thread
t_e = Thickness of external thread at critical shear plane
n = Threads per unit length

The gap equation is based on the tolerance data:

$E_{s,min}$ = Minimum pitch diameter of the external thread

$$t_e = 2 \cdot \frac{1}{\sqrt{3}}\left[\frac{3}{4}\frac{\sqrt{3}}{2n} - \frac{1}{2}\left(K_{n,max} + \frac{1}{2}\frac{\sqrt{3}}{2n} - E_{s,min}\right)\right]$$

$$t_e = \frac{1}{2n} + \frac{1}{\sqrt{3}}\left(E_{s,min} - K_{n,max}\right)$$

This equation is in the British Standards (BS) and American National Standards Institute (ANSI) standards and gives the shear area per unit length of engagement, L_e, to obtain the actual shear area.

$$A_{s,c} = \pi \cdot n \cdot K_{n,max}\left[\frac{1}{2n} + \frac{1}{\sqrt{3}}\left(E_{s,min} - K_{n,max}\right)\right]L_e$$

4.8.3 SHEAR AREA OF INTERNAL THREADS

$$A_{s,i} = \pi \cdot D_{s,min} \cdot t_i \cdot n$$

where:

$D_{s,min}$ = Minimum major diameter (external thread)
t_i = Thickness of internal thread (critical plane)

Similar to the previous derivation, the shear area of an internal thread (Figure 4.7) can be derived by taking into account the tolerances of the thread system.

$$A_{s,i} = \pi \cdot D_{s,min} \cdot n \cdot \left[\frac{1}{2n} + \frac{1}{\sqrt{3}}\left(D_{s,min} - E_{n,max}\right)\right]$$

$E_{n,max}$ = Maximum pitch diameter of the internal thread

4.9 LENGTH OF ENGAGEMENT

The length of engagement is considered the number of threads in engagement in either a nut or component multiplied by the thread pitch.

4.9.1 LENGTH OF ENGAGEMENT USING EQUAL STRENGTH MATERIALS

If the external and internal thread materials have the same strengths, then:

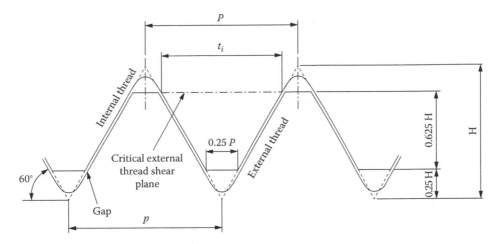

FIGURE 4.7   Shear area of internal thread.

Tensile strength (external thread)

$$S_t = \frac{F_{max}}{A_t}$$

Shear strength (internal thread)

$$0.5_{St} = \frac{F_{max}}{A_{s,i} \cdot L_e}$$

$$F_{max} = S_t A_t = 0.5 S_t \cdot A_{s,i}, L_e$$

Hence:

$$L_e = \frac{2A_t}{A_{s,i}}$$

4.9.2 Length of Engagement Using Dissimilar Strength Materials

If the external and internal thread material does not have the same strengths, then:

Tensile strength (external thread)

$$S_{t,e} = \frac{F_{max}}{A_t}$$

Shear strength (internal thread)

$$0.5_{t,i} = \frac{F_{max}}{A_{s,i} \cdot L_e}$$

$$F_{max} = S_{te} A_t = 0.5 S_{t,i} \cdot A_{s,i}, L_e$$

Hence:

$$L_e = \frac{2A_t \cdot S_{t,e}}{A_{s,i} \cdot S_{t,i}}$$

4.10   FASTENER AND NUT DESIGN PHILOSOPHIES

Standard fasteners and nuts of equal strengths are designed for the bolt to fail before the threads in the nut are stripped.

The design engineer has a responsibility to ensure that a failure of a machine element or other artifact will not endanger life. In this instance the length of thread engagement is an important consideration when designing machine elements using threaded fasteners.

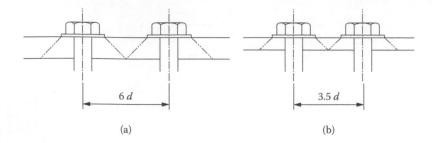

FIGURE 4.8   Effects of flange thickness on fastener pitch.

4.11   PITCHING OF FASTENERS

Considerable latitude exists for selecting the number of fasteners for a given application, as this will be influenced by the combination of pitch and size of fasteners in relation to the stiffness of the component being clamped.

Minimum pitch will be determined by spanner or socket clearances.

Maximum pitch may be determined by:

- Fretting.
- Leakage aspect. This will be determined by the permissible degree of sealing required: the type of media involved (i.e., oil, fuel air, etc.) and the pressures involved.

4.11.1   PRESSURE CONE

In general it is assumed that a 90° frustum of a cone commences at the edges of a spreader washer or fastener head. The aim is to have the base of each cone at least coincidental at the joint face. The effect of the flange thickness on the fastener pitch is shown in Figure 4.8.

An increase in the number of small-diameter fasteners is preferable to a lesser number of large-diameter fasteners. Where a small number of fasteners will adequately deal with the externally applied load, it is still desirable to increase the number to give a reasonable pitch. This will reduce the frettage effects, especially in the case of flanged bearing outer races, where, in the interest of weight saving, a small number would otherwise be adopted.

4.12   TENSION CONNECTIONS

The purpose of a bolt is to join two elements together. In general most bolted static joints when torqued to a particular value will be sufficient to maintain adequate joint pressures and meet its requirements.

It is important for the design engineer to ensure that the joint pressures are maintained through the life of the component and not lead to premature failure of the fastener due to either:

1. A significant overload causing a joint separation resulting in an overload in the fastener
2. The joint being subject to fatigue loading

This section will describe the elements of the bolted joint and lead to an understanding of a pretension diagram.

It is important to realize that threaded fasteners are designed for static tension loading. They will tolerate a small amount of shear loading together with a degree of bending in the head if the clamping surfaces are not truly parallel.

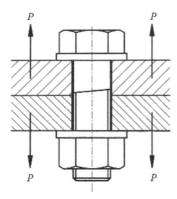

FIGURE 4.9 A typical bolted joint.

If the fastener is expected to take a large amount of shear load then shear bolts should be considered.

Alternatively, dowel pins should be used to take out the shear component in the fittings.

In Figure 4.9 a typical bolted joint is depicted with an external force P applied trying to separate the joint faces.

4.13 TORQUE-TENSION RELATIONSHIP

The formula (Equation (4.1)) is a good theoretical approximation for the torque required to tighten a nut onto a fastener, so to develop an end load P_i

$$T = P_i\left(\frac{p}{2\pi}\right) + \frac{E_s\mu_1}{2\ \cos\alpha} + \frac{d_o + d_i}{4}\mu_2 \tag{4.1}$$

The torque is in three parts: Pi(p/2π) represents that part of the torque absorbed in driving the mating thread helices over each other against the action of the axial load Pi to which they inclined, while $P_i(E_s\mu_1)/(2\ \cos\alpha)$ and $P_i((d_o - d_i)/4)\mu_2$ represent that absorbed in overcoming friction, in the first case between the threads and in the second case at the bearing face under the nut. A typical distribution between the terms would be, in the order of 10%, 40%, and 50% of the total. Friction conditions are therefore of predominating importance and unless known with a good degree of accuracy together with the bearing areas there is no point in using an expression more complicated than the simple empirical formula (Equation 4.2).

$$T = \frac{1}{5}P_i \cdot D \qquad \text{or} \qquad P_i = \frac{5T}{D} \tag{4.2}$$

where:
   T = Tightening torque
   P_i = Initial tensile load in the fastener developed by tightening
   D = Basic major diameter (i.e., nominal diameter)

This applies to nuts and bolts of normal proportions and for threads that are lubricated with a thin film of grease or oil. This is generally accurate to about ± 20%.

This does not apply to assemblies that are torque tightened from the bolt head.

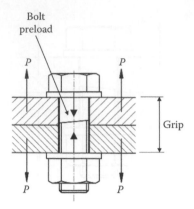

FIGURE 4.10   Bolt preload.

4.14   PROOF LOAD AND PROOF STRESS

4.14.1   FASTENER PRELOAD

A preload in a bolt-nut combination is achieved by tightening the nut to a specified torque setting (Figure 4.10).

It should be borne in mind that not all the torque applied to the nut will be converted into a load in the bolt. A substantial amount of torque will be lost in friction between the contact faces of the bolt head and the nut, together with another loss due to the friction within the thread itself.

There will be an extension in the bolt due to the preload and a small compressive force in the abutment. In the analysis of the fastener preload the bolt and its associated abutments are treated as springs.

The bolt stiffness is treated as two separated parts.

L_d = Plain portion of the bolt
L_t = Threaded portion of the bolt

This latter part is for the part of the thread above the nut. The bolt stiffness can be described (see Figure 4.11):

$$\delta = \delta_d + \delta_t$$

$$\delta_d = \frac{F_i}{k_d} d$$

$$\delta_t = \frac{F_i}{k_t} \tag{4.3}$$

$$F_i = \frac{k_d \, k_t}{k_d + k_t} \delta$$

$$F_i = k_b \cdot n$$

4.15   INTRODUCTION TO PRETENSION

It is not widely understood how a fastener in a fully tightened joint can survive in a situation where an untightened or loosely tightened fastener will fail within a very short space of time. When a

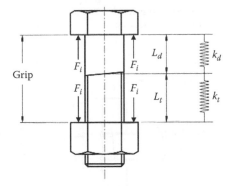

FIGURE 4.11 Bolt stiffness.

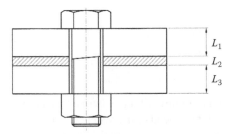

FIGURE 4.12 Abutment stiffness.

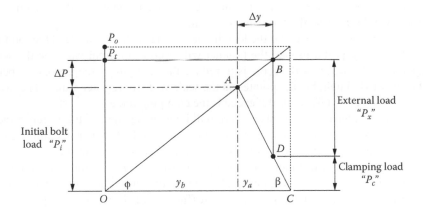

FIGURE 4.13 Tension diagram.

force is applied to a joint fitted with a fully tightened fastener, the fastener is exposed to only a small amount of the force. When a force is applied to a joint fitted with a fully tightened fastener as shown in Figure 4.9, the fastener is exposed to only a small amount of the force. The following description will explain this. The bolt can be treated as a solid spring as shown in Figure 4.11 and when fitted through the connecting parts as shown in Figure 4.12, the bolt is extended under the tightening force. The connecting parts are subjected to a compressive force under the abutment faces of the bolt head and nut. If these forces are plotted in a tension diagram (Figure 4.13), an understanding of the loads involved becomes obvious. The following text explains this in more detail.

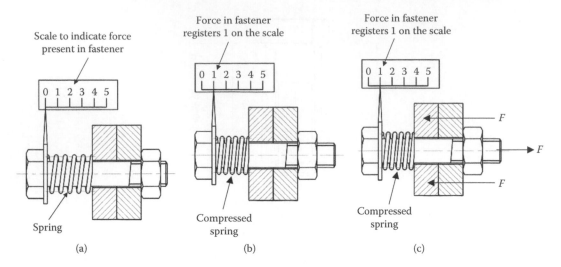

FIGURE 4.14 Description of preload.

4.15.1 WHY PRELOAD?

Consider the arrangement shown in Figure 4.14(a), where a fastener is located in a two-part joint and a compression spring is fitted under the head of the fastener; a pointer is also fitted to the head against a scale. The fastener has been partially tightened with a small load in the fastener and joint faces.

In Figure 4.14(b) the nut has been tightened to compress the spring so that the pointer indicates 1 on the scale. A working load has yet to be applied.

In the next figure (Figure 4.14(c)), with the same force in the fastener, a load of 1 unit is applied to the joint. This should have the effect of forcing the joint open, and it would be imagined that this additional force would increase the force in the fastener.

The initial reaction is to believe that the load in the fastener will increase and the joint faces will separate, producing a gap. But in fact, the fastener will keep its original value on the scale. What actually happens is the action of the force is to reduce the clamping force that exists between the end of the fastener to 0.5 unit, but the pointer on the scale will still register 1 unit. The fastener will not "feel" any of the applied force until it exceeds the clamping force.

Practically, when tightening takes place there will be an elongation of the fastener and a corresponding compression on the joint. This compression results in the fastener sustaining a proportion of the load. As the applied force reduces the clamping force that exists in the joint, an additional strain is felt by the fastener, which increases the force it sustains. The amount of the additional force the fastener sustains is less than the applied load acting on the joint. The actual amount of the force the fastener sustains will depend upon the ratio of stiffness of the fastener to the joint material.

4.16 JOINT DIAGRAMS

To help the design engineer visualize the loading within the bolted connection, joint diagrams have been developed. A joint diagram is a ready means of visualizing the load deflection characteristics of the fastener and the material it clamps. These diagrams assist in helping to understand how the joint sustains an external force and why the fastener does not sustain the whole of this force.

Figure 4.15 shows how a basic joint diagram is constructed. As the nut is tightened against the joint, the bolt will extend. Internal forces within the fastener resist this extension and a tension force or preload is generated. The reaction to this force is a clamping force that is the cause for the joint to be compressed. The force-extension diagram shows the fastener extension and the joint

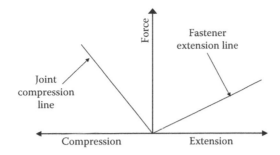

FIGURE 4.15 Basic joint diagram.

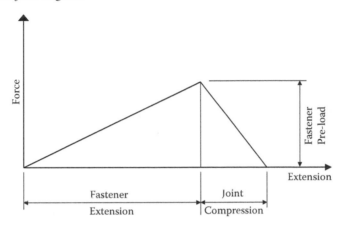

FIGURE 4.16 Basic pre-tension diagram.

compression. The slopes of the lines represent the stiffness of each component, the clamped joint being usually stiffer than the fastener.

Moving the compression line of the joint to the right forming a triangle, as it is recognized, forms the basic diagram shown in Figure 4.16, with the fastener and the joint in equilibrium. The clamping force tending to compress the joint is then seen to be equal to the preload in the fastener. Positive extension is to the right, such as that sustained by the fastener; negative extension (compression) is to the left and is sustained by the joint material.

4.16.1 JOINT DIAGRAMS WITH AN EXTERNAL LOAD APPLIED

When an external force is applied to the joint it has the effect of reducing some of the clamping force caused by the fastener preload and applying an additional force to the fastener itself. This is illustrated in the joint diagram in Figure 4.17.

The external force acts through the joint material and then subsequently into the fastener.

It can be seen that the load on the fastener cannot be added without a subsequent reduction in the clamp force acting on the joint. As can be observed from a study of the diagram, the actual amount of increase in the fastener force is dependent upon the relative stiffness of the fastener to the joint.

As an illustration of the importance of the relative stiffness of the fastener to the joint, two figures are shown, Figure 4.18(a) presents a hard joint (low stiffness fastener with a high stiffness joint) and Figure 4.18(b) a soft joint (a high stiffness fastener with a low stiffness joint).

With a hard joint, because of the steep joint stiffness slope, in this case the fastener will only sustain a small proportion of the applied force.

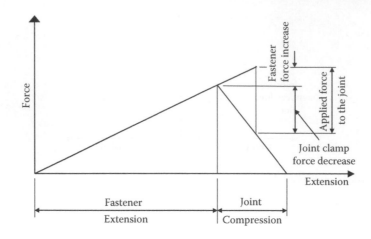

FIGURE 4.17 Pre-tension diagram with preload.

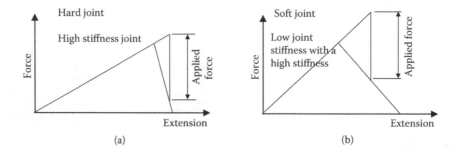

FIGURE 4.18 Effects of fastener stiffness.

A soft joint has a slope for the fastener greater than that of the joint. The fastener will therefore sustain the majority of the applied force.

High-performance fasteners generally have their shank diameters reduced down to the stress diameter of the thread. If the fastener is modified in this manner the stiffness of the fastener is reduced so that it will not support as much of any applied load that it would otherwise do. Provided that the shank diameter is not reduced below that of the stress diameter, the strength of the fastener will not normally be impaired.

4.16.2 Effects of a Large Increase in the External Load

With an increase in the external load, the force acting on the fastener proportionally increases, and at the same time there is a reduction in the clamping force acting on the joint.

If the external force continues to increase, then either:

1. The proportion of the external force acting on the fastener's preload would result in the yield of the fastener material being exceeded with the possibility of the fastener failure.

   Even if failure does not occur immediately, with the removal of the external load the preload will be reduced due to a permanent deformation of the fastener.

2. The clamping force acting on the joint will continue to reduce to a point when it becomes zero. Any further increase in the external force will result in a gap forming within the joint, with the fastener then supporting all the additional force (Figure 4.19).

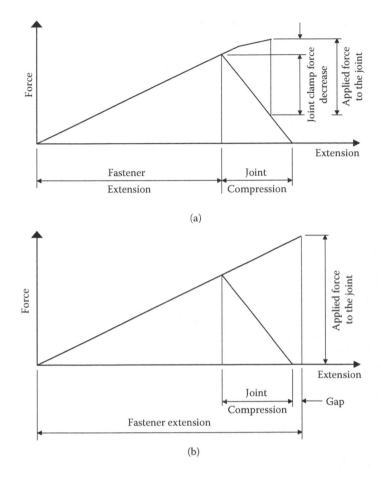

FIGURE 4.19   Effect of increase in the external load.

If a gap forms within the joint comprising the joint, then the fasteners are almost always subjected to nonlinear loadings from bending and shear forces. This will quickly result in the fastener failure; therefore, it is normal for the design criteria to include a statement that the applied forces must not under any circumstances result in a gap forming within the joint.

4.16.3   THE EFFECT OF A COMPRESSIVE EXTERNAL LOAD

If a joint is subjected to a compressive external force (Figure 4.20), this has the effect of increasing the clamping force acting on the joint and decreasing the tension within the fastener. If the compressive external force is large enough, then:

1. The tension in the fastener can be reduced to a low value; if the joint is subjected to a cyclic load, the fastener could fail due to fatigue as it experiences tension variations under a compressive force. The fastener will be most susceptible to loosening due to vibration.
2. The yield limitations of the clamped material are exceeded, as the joint is supporting a compressive load in addition to that provided by the fasteners' preload. A loss of preload will result from permanent deformation when the external load is released.

Bolt pretension, also called preload or prestress, is generated when a nut is tightened on a threaded bolt or screw. The load in the fastener increases and the deformation of the fastener also increases.

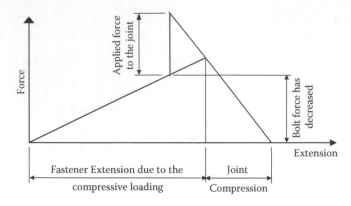

FIGURE 4.20   Effect of a compressive load.

4.16.4   NOMENCLATURE

A = Area of plate subject to load
A_b = Bolt cross section (bolt thread root area) (mm^2)
D_b = Bolt thread root diameter (mm)
d = Bolt nominal diameter (mm)
F_e = External load (N)
F_t = Thermal load (N)
F_p = Preload (N)
F = Total load on bolt (N)
E_z = Modulus of elasticity (N/mm^2)
L = Length of bolt joint (mm)
L_b = Length of bolt (mm)
L_j = Length of joint (mm)
k_z = Stiffness of component (N/mm)
k_b = Stiffness of bolt (N/mm)
k_j = Stiffness of joint (N/mm)
t_z = Thickness of plate (mm)
T = Bolt tightening torque (Nmm)
x_z = Deflection of item z/unit load (mm/N)
x_b = Deflection of bolt/unit load (mm/N)
α_z = Coefficient of thermal expansion of component z (mm/mm/deg.C)
δ = Deflection (mm)

4.16.5   NOTES

A bolt tightened to a specific value is considered safer than a bolt simply tightened to an arbitrary value. A preload between 75 and 80% of the proof strength of the bolt material is normally used.

Consider a bolt being used to clamp a joint to a set preload value. The bolt has a low stiffness and the joint has a very high stiffness. An external load, when applied to the joint, will tend to separate the joint. Part of this load will cause a further extension in the bolt, and the remaining part of the load will result in a reduction of the compressive load on the joint.

It follows that an infinitely stiff bolt will result in no separation of the joint faces. The question is to determine a realistic stiffness for the bolt to withstand the influences of the external force.

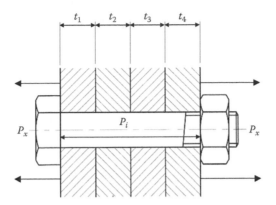

FIGURE 4.21 Bolted assembly with a preload and external load applied.

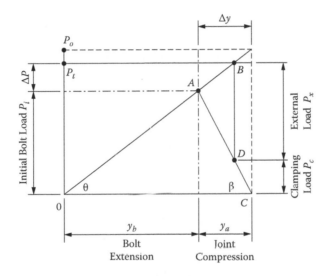

FIGURE 4.22 Bolt loading diagram with preload and external load applied.

Consider Figure 4.21, which shows an assembly bolted with a preload of P_i and an external load of P_x is applied.

For an infinitely stiff bolt, separation will not occur, as all the external load will be directly applied to the bolt with no resulting extension. For an infinitely stiff joint, separation will take place when the external load exceeds the preload.

The bolt loading diagrams (Figures 4.22 and 4.23) show the loading regime on the bolt and joint.

The calculation for the proportion of the load taken by the bolt and the joint using the component stiffness values uses the same basic formula as that for spring rate, i.e.,

$$\text{Stiffness} = k = \frac{P}{\delta}$$

4.17 FASTENER STIFFNESS

The relationship $E = \text{stress/strain} = f/\varepsilon$ is used to determine the stiffness of a bolt.

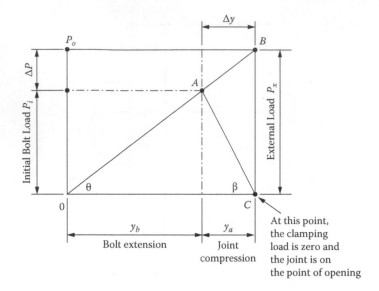

FIGURE 4.23 Bolt loading diagram with external load matching preload.

$$\text{Stress (f)} = \frac{\text{Force}}{\text{Area}} = \frac{P}{A}$$

$$\text{Strain (e)} = \frac{\text{Deflection}}{\text{Length}} = \frac{d}{L}$$

$$E = \frac{\dfrac{P}{A}}{\dfrac{\delta}{L}}$$

Therefore

$$\delta = \frac{P \cdot L}{E \cdot A}$$

$$k_b = \frac{P_b}{\delta_b} = \frac{P_b}{\dfrac{P_b \cdot L_b}{A_b \cdot E_b}} = \frac{A_b \cdot E_b}{L_b}$$

If the bolt length clamping the joint includes a number of different sections, then the resulting stiffness is determined using the relationship

$$\frac{1}{k_{bt}} = \frac{1}{k_1} + \frac{1}{k_2} + \frac{1}{k_3} + \frac{1}{k_4}$$

To allow for a certain degree of elasticity of the bolt head and nut, a correction factor is often used to modify the length used in the stiffness calculations shown in Figure 4.24.

The stiffness of the bolt results from the stiffness of the bolt shank (diameter d_s) and the stiffness of the bolt thread (root diameter d_r).

The length used to calculate the shank stiffness = $L_{se} = L_s + 0.4 d_s$.

The length used for the threaded length section = $L_{te} + L_t + 0.4 d_r$.

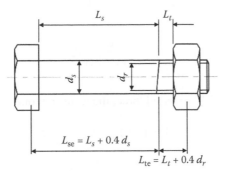

FIGURE 4.24   Stiffness characteristics of a bolt.

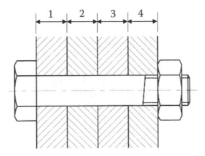

FIGURE 4.25   Multiple clamping surfaces.

4.18   JOINT STIFFNESS

It is very difficult to calculate the stiffness of a joint that is based on holes drilled in a plate (see Figure 4.25). A rough estimation can be made by assuming the joint is an annulus with an outside diameter (OD) of 2.5 times the bolt diameter and an inside diameter (ID) equal to the bolt diameter.

The total joint stiffness is related to the individual stiffness values as follows:

$$\frac{1}{k_{jt}} = \frac{1}{k_1} + \frac{1}{k_2} + \frac{1}{k_3} + \frac{1}{k_4}$$

4.18.1   CALCULATION OF LOAD DISTRIBUTION USING FASTENER/JOINT STIFFNESS

A joint preload with a force P_x is then subject to an additional load P_e, which tends to separate the joint. The resulting deflections of the joint and bolt are the same providing that P_e is less than the separating force.

$$d = \frac{F_{be}}{k_b} = \frac{F_{je}}{k_j} \qquad F_e = F_{be} + F_{je}$$

It follows that

$$F_{be} = F_{je} \frac{k_b}{k_j}$$

$$= \frac{k_b}{k_j} \left(F_e - F_{be} \right)$$

$$= \frac{k_b}{\left(k_b + k_j \right)} F_e$$

Also

$$F_{je} = \frac{k_j}{\left(k_b + k_j\right)} F_e$$

Following the application of the external force, the resulting total force on the bolt is

$$F = F_p + F_{be}$$

$$= F_p + \frac{k_b}{\left(k_b + k_j\right)} F_e$$

the total force on the joint

$$F_j = F_p + F_{je}$$

$$= \frac{k_j}{\left(k_b + k_j\right)} F_j - F_p$$

4.19   THERMAL LOADING

If all the materials of the joint and the bolt are the same, then any changes in temperature will have minimal effects on the joint loadings. However, if the joint materials have thermal expansion coefficients different than the bolt material, changes in the joint loading will result from changes in temperature.

Coefficient of thermal expansion of joint material = α_j

Coefficient of thermal expansion of bolt = α_b

Change of temperature = ΔT

Length of joint = length of bolt = $\left(L_j = L_b = L\right)$
(as ambient temperature)

Expansion of joint = $\Delta L_j = \alpha_j \Delta T \cdot L$

Expansion of bolt = $\Delta L_b = \alpha_b \Delta T \cdot L$

Overall stiffness of the joint is calculated as follows:

$$\frac{1}{k_t} = \frac{1}{k_j} + \frac{1}{k_b}$$

Therefore

$$k_t = \frac{k_b \cdot k_j}{\left(k_b + k_j\right)}$$

The resulting change in the joint load is calculated as follows:

$$F_t = \frac{k_b \cdot k_j}{\left(k_b + k_j\right)}\left(\Delta L_j - \Delta L_b\right)$$

$$= \frac{k_b \cdot k_j}{\left(k_b + k_j\right)}\Delta T \cdot L\left(\alpha_j - \alpha_b\right)$$

The total bolt load following temperature change:

Let
$$F = F_p + F_t$$

4.19.1   INITIAL TENSION IN BOLT

The initial tension in a bolt is roughly estimated for a bolt tightened by hand of an experienced mechanic, as follows.

The tension resulting from this equation would be reasonably safe for M8 grade 8.8 bolts and above.

$$F_p = K \cdot d$$

where:

F_p = Preload (N)
K = Coefficient (varies between 1.75×106 N/m and 2.8×106 N/m)
d = Nominal diameter of bolt (mm)

For a bolt tightened with a torque wrench, the torque required to provide an initial bolt tension may be approximated by the formula

$$T = F_p \cdot K \cdot d$$

Typical K factors:

Steel thread condition	K
As received, stainless on mild or alloy	0.30
As received, mild or alloy on same	0.20
Cadmium plated	0.16
Molybdenum-disulfide grease	0.14
Polytetrafluoroethylene (PTFE) lubrication	0.12

A more accurate value can be obtained from the following:

$$T = \text{thread torque} + \text{collar torque}$$

$$= \frac{F_p \cdot d_m}{2}\left(\frac{p + \pi \cdot \mu \cdot d_m \cdot \sec\ \alpha}{\pi \cdot d_m - \mu \cdot p \cdot \sec\ \alpha}\right) + F_p \cdot \mu_c \cdot r_c$$

where
F_p = desired bolt preload (N)
p = thread pitch (mm)
d_m = mean diameter of thread (mm)
μ = coefficient of thread friction
μ_c = coefficient of collar friction

α = the thread angle/2 ($\alpha = 30°$ for standard metric threads and $\alpha = 29°/2$ for acme threads).
r_c = collar friction radius (mm)

It can be proved that the majority of the torque is required to overcome the thread and collar friction forces (approximately 90%). Therefore, any errors in the value of the friction coefficient will have a large variation on the bolt tensile load. The above formula is in essence not a lot more accurate than the approximate formulas above.

4.20 FASTENERS SUBJECT TO COMBINED SHEAR AND TENSION

4.20.1 INTERACTION CURVES: LOAD RATIOS AND FACTORS OF SAFETY

Bolts and screws under combined tension stresses or loads may be predicted without using principal stresses or loads by using the interactive method.

The method represents applied and allowable stress or load conditions on the shank of a fastener by loads.

Load ratios are nondimensional coefficients R given by

$$R = \frac{\text{applied load}}{\text{allowable load}}$$

The method involves determining the allowable load for each separate failure mode such as tension, compression, bending, buckling, and shear. Load ratio R for each separate failure condition is calculated and combined using interaction equations if these loads act simultaneously on the fastener.

The equations generally take the following form:

$$R_1^x + R_2^y + R_3^z \ldots = 1.0$$

where ^{xyz} are exponents defining interaction relationship.

Failure is when the sum of the load ratios is greater than 1.0.

4.20.2 INTERACTION CURVE

The margin of safety (MoS) assuming each load increases proportionately until failure occurs at a point B (see Figure 4.26):

$$MS = \frac{OB}{OA} - 1$$

The margin of safety assuming R_1 remains constant with R_2 increasing until failure occurs at point C:

$$MS = \frac{EC}{EA} - 1$$

The margin of safety assuming R_2 remains constant with R_1 increasing until failure occurs at point D:

$$MS = \frac{FD}{FA} - 1$$

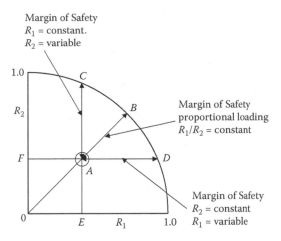

Point A is located with the coordinates R_1 and R_2

FIGURE 4.26 Interaction curve.

4.20.3 INTERACTION EQUATION

$$MS = \frac{1}{\sqrt{R_t^2 + R_s^2}} - 1$$

where:

$$R_t = \frac{\text{applied ultimate tension load}}{\text{allowable ultimate tension load}}$$

$$R_s = \frac{\text{applied ultimate shear load}}{\text{allowable ultimate shear load}}$$

Note that when calculating R_s, shear is applied to the bolt shank; this area should be used in calculating the allowable shear load.

4.21 ECCENTRIC LOADS

The design engineer is sometimes confronted with a problem requiring the design of a bracket to attach to an existing structural member, such as that depicted in Figure 4.27.

In this case the line of action passes outside the bolt group and as a consequence will create a shear load on the fasteners.

Figure 4.28 considers the loading seen by each fastener.

Let u_1 = load/fastener per unit distance from O.

$$PL = 4u_1 I_a^2 + 2u_1 I_b^2 \qquad (4.3)$$

EXAMPLE 4.1

Consider a bracket similar in design to that depicted in Figure 4.28 with the principal load and dimensions as follows:

P = 150 kN
L = 225 mm
I_a = 111.803 mm
I_b = 111.803 mm
n = 6

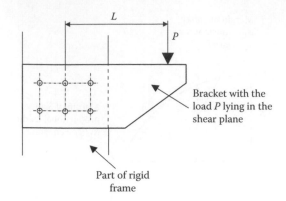

FIGURE 4.27   Bracket subject to shear loading.

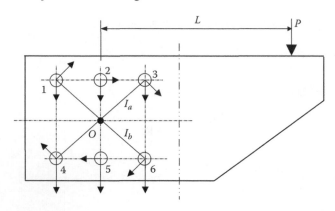

FIGURE 4.28   Centroid of bolt group.

Determine the size of the fasteners if the allowable shear stress on the fasteners is to be limited to 400 MPa.

Rearranging Equation (4.3) to solve for u_1:

$$u_1 = \frac{P.L}{\left[\left(4J_a{}^2 \right) + \left(2J_b{}^2 \right) \right]}$$

$$u_1 = 450.0 \ \frac{kN}{m}$$

Load on fastener due to turning moment:

$$L_t = u_1.J_a$$

$$L_t = 50.312 \ kN$$

Now, direct load on fastener due to load:

$$L_d = \frac{P}{n}$$

$$L_d = 250 \ kN$$

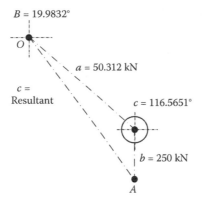

FIGURE 4.29 Angle of forces for Example 4.1.

Resultant load (solve using the sine and cosine rules) (see Figure 4.29 for angle of forces):

$$c = \frac{L_t \cdot 0.894426}{0.6877428}$$

$$c = 65.432 \, kN$$

This is the maximum shear load acting on fastener 6.

4.21.1 PERMISSIBLE SHEAR STRESS

$$f_{fastener} = 400 \, MPa$$

Diameter of fastener:

$$dia' \, (d) = \sqrt{\frac{4 \cdot f_s}{\pi \cdot f_{fastener}}}$$

$$d = \sqrt{\frac{4 \cdot 65.432 \, kN / mm^2}{\pi \cdot 400 \, MPa}}$$

$$= 14.432 \, mm$$

Selecting a stock size fastener:

$$d = M16 \times 1.75 \, pitch$$

4.22 PRYING FORCES

EXAMPLE 4.2

Consider the bracket in Figure 4.30 where the applied force is in the plane of the connections. In this case the load produces a tension and shear force in the fasteners. (The fasteners have no preload applied in this example.)

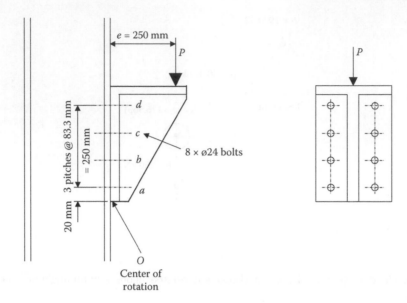

FIGURE 4.30 Bracket without preload.

Maximum stress

$$f_{tension} = \frac{F_t}{A}$$

$$f_{tension} = 61.38 \text{ MPa}$$

$$f_{shear} = \frac{P}{n.A}$$

$$f_{shear} = 59.683 \text{ MPa}$$

$P = 150 \text{ kN}$ $y_a = 20 \text{ mm} + y$

$e = 250 \text{ mm (Moment arm)}$ $y_b = 20 \text{ mm} + (2 \times y)$

$y = 83.3 \text{ mm pitch}$ $y_c = 20 \text{ mm} + 3 \times y)$

$Y = 250 \text{ mm}$ $y_d = 20 \text{ mm} + (4 \times y)$

$n = 8 \text{ bolts}$

$d = 20 \text{ mm}$

$A = \dfrac{\pi \cdot d^2}{4}$ $\Sigma y^2 = (y_a^2 + y_b^2 + y_c^2 + y_d^2)$

$A = 314.16 \text{ mm}^2$ $\Sigma y^2 = 243.087 \times 10^3 \text{ mm}^2$

$P_1 = P \cdot e$

$P_1 = 37.5 \times 10^3 \text{ kN} / \text{mm}^2$

$Y_1 = \dfrac{\Sigma y^2}{Y}$

$Y_1 = 972.347 \text{ mm}$

The resultant stress on the top fastener is a vector quantity derived from the tensile stress $f_{tension}$ and the direct shear stress f_{shear}.

$$F_r = (f_{tension}^2 + f_{shear}^2)^{0.5}$$

$$F_r = 85.613 \text{ MPa}$$

Using fasteners to BS3692 strength grade 4.6, yield strength = 235 kN/mm².

$$f_{allowable} = 235 \text{ MPa}$$

Factor of Safety (FoS)

$$FoS = \frac{f_{allowable}}{F_r}$$

$$FoS = 2.745$$

Tensile load on top fastener.

$$F_t = \frac{P \cdot e}{2 \cdot Y_1}$$

$$F_t = 19.283 \text{kN}$$

Direct load

$$F_s = \frac{P}{n}$$

$$F_s = 18.75 \text{kN}$$

Note that the above calculation is based on the fact that due to no preload on the fasteners the bracket will heel about the point of origin O. It is considered that if the prying load is high enough, plastic deformation will occur in the bracket at this point.

If this happens, then this will have the effect of reducing the loads, as the heel point will migrate upwards toward the first row of fasteners. Some argue that this is the point where the moment should be measured from. It is considered that the former position should be adopted, as the second position will have the effect of reducing the load on the fasteners.

In this second example of the above bracket (Figure 4.31), the fasteners have preload applied, so this will minimize any separation between the bracket and the supporting steelwork. From Figure 4.30 it is clearly seen that the fastener deflections due to the applied moment are significantly reduced compared to Example 4.1.

4.23 FASTENERS SUBJECT TO ALTERNATING EXTERNAL FORCE

Consider a M12 × 1.75p grade 8.8 bolt used to clamp a fitting to a structural member.

The cross section of the part is (12 × 16 mm) = 192 mm² and is subject to an alternating load between 0 and 20,000 N.

Assume the stress concentration factor (SCF) for the thread is 3.0. Calculate the following:

1. The FoS for the bolt without any preload
2. The minimum preload required to prevent any loss of compression
3. The FoS for the bolt with a preload of 22,000 N
4. The minimum force in the part when the preload is 22,000 N

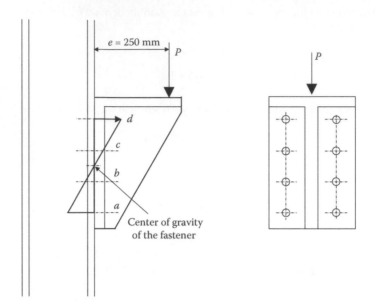

FIGURE 4.31   Bracket with preload.

From BS 6104, Part 1, 1981, Table 6, the ultimate tensile load for this size and type of fastener is quoted as 67,400 N. The fastener core area is 84.267 mm².

The ultimate tensile stress (UTS) will be (67,400 N/84.267 mm²) = 799.838 MPa. It is assumed that the yield strength is 500 MPa and the fatigue endurance limit for this material is 50% of the UTS = 400 MPa.

4.23.1   Factor of Safety (FoS) with No Preload

The factor of safety (FoS) is determined by multiplying the mean stress and the stress amplitude, multiplied by the SCF and plotted on a modified Goodman diagram.

The mean stress
$$S_m = \frac{20000}{2 \times 84.3} = 118.6 \text{ MPa}$$

With no preload the entire load will be carried by the fastener.

For a load alternating between 0 and 200,000 N, the mean load is 10,000 N and the amplitude of the stress, multiplied by the SC (3.0), is (118.6 MPa × 3) = 355.8 MPa.

Plotting this on the diagram in Figure 4.32 shows that point A falls outside the safe zone, and therefore the fastener will be unsafe.

$$\frac{1}{\text{FoS}} = \left(\frac{\text{mean stress}}{S_y} \right) + \left(\frac{\text{stress amplitude} \times \text{scf}}{S_e} \right)$$

$$\frac{1}{\text{FoS}} = \left(\frac{118.6}{500} \right) + \left(\frac{118.6 \times 3.0}{400} \right)$$

hence:
$$\text{FoS} = 0.887$$

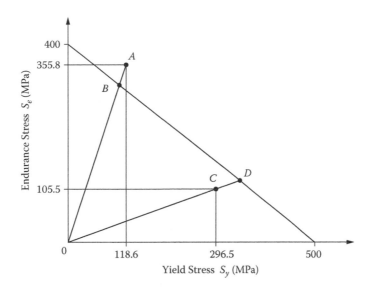

FIGURE 4.32 Modified Goodman diagram for Example 4.3.

4.23.2 THE MINIMUM PRELOAD TO PREVENT ANY LOSS OF COMPRESSION

For components of similar material and equal length, the spring constants are proportional to the cross-sectional areas of the part and fastener.

Therefore, when the part has zero compression, the preload in the fastener is

$$P_{\text{pre-load}} = \frac{A_s \cdot P_{\text{max}}}{\left(A_s + A_p\right)}$$

$$= \frac{192.0 \ \text{mm}^2 \times 20000 \ \text{N}}{\left(84.267 \ \text{mm}^2 + 192.0 \ \text{mm}^2\right)}$$

$$= 13.90 \ \text{kN}$$

4.23.3 CALCULATE THE FOS FOR THE BOLT WITH A PRELOAD OF 22,000 N

$$P_{\text{b average}} = \frac{P_{\text{average}} \times A_b}{(A_p + A_b)} + P_{\text{pre-load}}$$

$$P_{\text{b average}} = \frac{10000 \ \text{N} \times 84.267 \ \text{mm}^2}{192.0 \ \text{mm}^2 + 84.267 \ \text{mm}^2} + 22.00 \ \text{kN}$$

$$= 25050.2 \ \text{N}$$

$$\text{Stress}_{\text{mean}} = \frac{P_{\text{b average}}}{A_b}$$

$$= \frac{25050.2 \text{ N}}{84.267 \text{ mm}^2}$$

$$= 297.27 \text{ MPa}$$

$$P_{\text{b amplitude}} = \frac{A_b \times P_{\text{b average}}}{A_b + A_p}$$

$$= \frac{84.267 \text{ mm}^2 \times 10000 \text{ N}}{84.267 \text{ N} + 192.0 \text{ N}}$$

$$= 3.0502 \text{ kN}$$

$$\text{Stress}_{\text{amplitude}} = \frac{P_{\text{b amplitude}}}{A_b}$$

$$= \frac{3050.2 \text{ N}}{84.267 \text{ mm}^2}$$

$$= 36.197 \text{ MPa}$$

Point C (Figure 4.32) is inside the safe zone when plotted on the diagram; therefore, the bolt has a FoS when measured graphically from the diagram OD/OC:

$$\frac{1}{\text{FoS}} = \frac{\text{OC}}{\text{OD}} = \frac{314.29}{368.51}$$

$$\frac{1}{\text{FoS}} = 0.8529$$

$$\text{FoS} = 1.173$$

Algebraically:

$$\frac{1}{\text{FoS}} = \left(\frac{297.27 \text{ MPa}}{500.0 \text{ MPa}} \right) + \left(\frac{3.0 \times 36.197 \text{ MPa}}{400.0 \text{ MPa}} \right)$$

$$\text{FoS} = 1.155$$

The effect of the preload is to significantly reduce the magnitude of the alternating force and stress in the fastener (which is normally much more critical) while increasing the mean fastener force and stress (normally less critical). This is shown in the lower part of Figure 4.33.

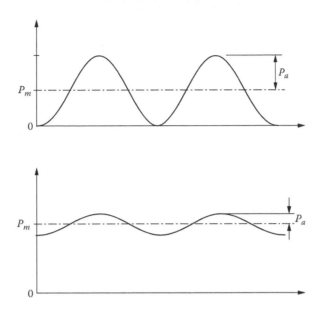

FIGURE 4.33   Effect of preload on the magnitude of the alternating force.

4.23.4   THE MINIMUM FORCE IN THE PART WHEN THE PRELOAD IS 22,000 N

$$P_{\text{pre-load min}} = A_p \frac{P_{\max}}{A_b + A_b} - P_{\text{pre-load}}$$

$$P_{\text{pre-load min}} = \left(\frac{192.0 \text{ mm}^2 \times 20000 \text{ N}}{192.0 \text{ mm}^2 + 84.267 \text{ mm}^2} \right) - 22.00 \text{ kN}$$

$$= -81.00 \text{ kN} \quad \text{Compression}$$

5 Limits and Fits

5.1 INTRODUCTION

The International Organization for Standardization (ISO) system (ISO 286-1988 and 2:1988) (on which the British Standards BS EN 20286-1993 and 2:1993 are based) provides for a wide range of hole and shaft tolerances to cater for a wide range of conditions.

However, it has been found for normal manufacturing processes that the range can be condensed to a small number of tolerance combinations.

The most commonly applied hole and shaft tolerances include:

Selected hole tolerances: H7, H8, H9, and H11

Selected shaft tolerances: c11, d10, e9, f7, g6, h6, k6, n6 p6, s6 and u6

The general standard adopted is the basic hole; the measurement of holes is generally controlled using plug gauges with "go no go" features, whereas a shaft is easily checked using micrometers or gap gauges.

In some instances a manufacturer may choose to use a shaft-based system, particularly where a single shaft may have to accommodate a variety of accessories, such as bearings, couplings or collars, etc.

Shaft-based systems may offer economies particularly where bar stock material may be available to standard shaft tolerances to the ISO system. For the nomenclature used in limits and fits, see Figure 5.1.

5.2 TOLERANCE GRADE NUMBERS

5.2.1 TOLERANCE

Tolerance is the difference between the maximum and minimum size limits of a component.

5.2.2 INTERNATIONAL TOLERANCE GRADE NUMBERS

Tolerance grade numbers are used to specify the size of a tolerance zone.

In the British Standard, the tolerance is the same for both the internal (hole) and external (shaft) parts having the same tolerance grade numbers.

Tolerance grades IT0 through IT16 are covered in the National and International standards.

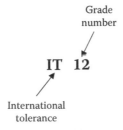

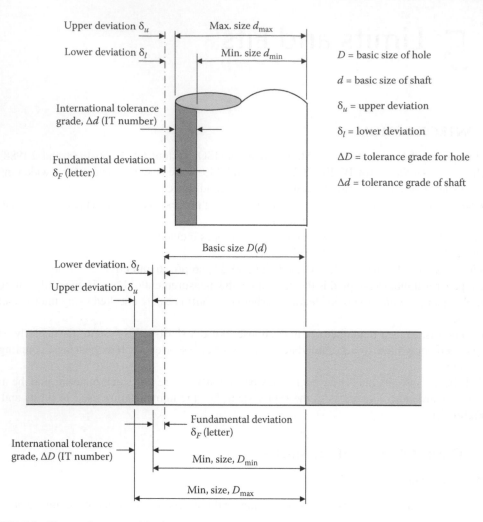

FIGURE 5.1 Nomenclature used in limits and fits.

5.3 FUNDAMENTAL DEVIATIONS

5.3.1 PREAMBLE

Consider a 25.000 mm nominal diameter hole and shaft.
    An example of a fit specification:

25H7 hole
25g6 shaft

Note:
  • All hole deviations are specified in uppercase.
  • All shaft deviations are specified in lowercase.

5.3.2 FUNDAMENTAL DEVIATION

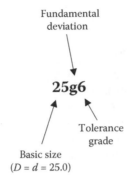

From the tolerance grade table for a basic size of 25.000 mm diameter, the variation between the upper and lower deviations for a grade IT6 will be 0.013 mm (see Table 5.1, 25.000 mm), and for grade IT7 will be 0.021 mm.

From Table 5.2, for ⌀25.000 g the upper deviation is –0.007 mm. The lower deviation will be –0.007 to 0.013 mm = 0.020 mm.

Hence the tolerance on a 25.000 mm diameter shaft machined to g6 is

⌀25.000 mm –0.007/–0.020 mm

TABLE 5.1
Tolerance Grades

Basic Sizes	Tolerance Grades							
	IT5	IT6	IT7	IT8	IT9	IT10	IT11	IT12
0–3	0.004	0.006	0.010	0.014	0.025	0.040	0.060	0.100
3–6	0.005	0.008	0.012	0.018	0.030	0.048	0.075	0.120
6–10	0.006	0.009	0.015	0.022	0.036	0.058	0.090	0.150
10–18	0.008	0.011	0.018	0.027	0.043	0.070	0.110	0.180
18–30	0.009	0.013	0.021	0.033	0.052	0.084	0.130	0.210
30–50	0.011	0.016	0.025	0.039	0.062	0.100	0.160	0.250
50–80	0.013	0.019	0.030	0.046	0.074	0.120	0.190	0.300
80–120	0.015	0.022	0.035	0.054	0.087	0.140	0.220	0.350
120–180	0.018	0.025	0.040	0.063	0.100	0.160	0.250	0.400
180–250	0.020	0.029	0.046	0.072	0.115	0.185	0.290	0.460
250–315	0.022	0.032	0.052	0.081	0.130	0.210	0.320	0.520
315–400	0.025	0.036	0.057	0.089	0.140	0.230	0.360	0.570
400–500	0.027	0.040	0.063	0.097	0.155	0.250	0.400	0.630

Source: BS EN 20286-2:1993.
Note: Values in mm

5.3.3 FUNDAMENTAL DEVIATIONS FOR SHAFTS

TABLE 5.2
Fundamental Deviation (Shaft)

Basic Size	Upper Deviation Letter					Lower Deviation Letter				
	c	d	f	g	h	k	n	p	s	u
0–3	–0.060	–0.020	–0.006	–0.002	0	0	0.004	0.006	0.014	0.018
3–6	–0.070	–0.030	–0.010	–0.004	0	0.001	0.008	0.012	0.019	0.023
6–10	–0.080	–0.040	–0.013	–0.005	0	0.001	0.010	0.015	0.023	0.028
10–14	–0.095	–0.050	–0.016	–0.006	0	0.001	0.012	0.018	0.028	0.033
14–18	–0.095	–0.050	–0.016	–0.006	0	0.001	0.012	0.018	0.028	0.033
18–24	–0.110	–0.065	–0.020	–0.007	0	0.002	0.015	0.022	0.035	0.041
24–30	–0.110	–0.065	–0.020	–0.007	0	0.002	0.015	0.022	0.035	0.048
30–40	–0.120	–0.080	–0.025	–0.009	0	0.002	0.017	0.026	0.043	0.060
40–50	–0.130	–0.080	–0.025	–0.009	0	0.002	0.017	0.026	0.043	0.070
50–65	–0.140	–0.100	–0.030	–0.010	0	0.002	0.020	0.032	0.053	0.087
65–80	–0.150	–0.100	–0.030	–0.010	0	0.002	0.020	0.032	0.059	0.102
80–100	–0.170	–0.120	–0.036	–0.012	0	0.003	0.023	0.037	0.071	0.124
100–120	–0.180	–0.120	–0.036	–0.012	0	0.003	0.023	0.037	0.079	0.144
120–140	–0.200	–0.145	–0.043	–0.014	0	0.003	0.027	0.043	0.092	0.170
140–160	–0.210	–0.145	–0.043	–0.014	0	0.003	0.027	0.043	0.100	0.190
160–180	–0.230	–0.145	–0.043	–0.014	0	0.003	0.027	0.043	0.108	0.210
180–200	–0.240	–0.170	–0.050	–0.015	0	0.004	0.031	0.050	0.122	0.236
200–225	–0.260	–0.170	–0.050	–0.015	0	0.004	0.031	0.050	0.130	0.258
225–250	–0.280	–0.170	–0.050	–0.015	0	0.004	0.031	0.050	0.140	0.284
250–280	–0.300	–0.190	–0.056	–0.017	0	0.004	0.034	0.056	0.158	0.315
280–315	–0.330	–0.190	–0.056	–0.017	0	0.004	0.034	0.056	0.170	0.350
315–355	–0.360	–0.210	–0.062	–0.018	0	0.004	0.037	0.062	0.190	0.390
355–400	–0.400	–0.210	–0.062	–0.018	0	0.004	0.037	0.062	0.208	0.435

Source: BS EN20286-2:1993.
Note: Values in mm

5.3.4 FUNDAMENTAL DEVIATIONS FOR HOLES

TABLE 5.3
Fundamental Deviation (Hole)

Basic Size	Lower Deviation H
0–3	0.000
3–6	0.000
6–10	0.000
10–14	0.000

TABLE 5.3 *(Continued)*
Fundamental Deviation (Hole)

Basic Size	Lower Deviation H
14–18	0.000
18–24	0.000
24–30	0.000
30–40	0.000
40–50	0.000
50–65	0.000
65–80	0.000
80–100	0.000
100–120	0.000
120–140	0.000
140–160	0.000
160–180	0.000
180–200	0.000
200–225	0.000
225–250	0.000
250–280	0.000
280–315	0.000
315–355	0.000
355–400	0.000

Note: Values in mm

5.3.5 UPPER AND LOWER DEVIATIONS

5.3.5.1 Shaft Letter Codes c, d, f, g, and h

Upper deviation is equivalent to the fundamental deviation.
    Lower deviation is equivalent to:

$$\text{Upper deviation} - \text{tolerance grade}$$

5.3.5.2 Shaft Letter Codes k, n, p, s, and u

Lower deviation is equivalent to the fundamental deviation.
    Upper deviation is equivalent to:

$$\text{Lower deviation} + \text{tolerance grade}$$

5.3.5.3 Hole Letter Code H

$$\text{Lower deviation} = 0$$

$$\text{Upper deviation} = \text{tolerance grade}$$

5.4   PREFERRED FITS USING THE BASIC HOLE SYSTEM

TABLE 5.4
Preferred Fits Using the Basic Hole

Type of Fit	Description	Symbol
Clearance	**Loose running fit:** For wide commercial tolerances or allowances on external members.	H11/c11
	Free running fit: Not for use where accuracy is essential, but good for large temperature variations, high running speeds, or heavy journal pressures.	H9/d9
	Close running fit: For running on accurate machines and for accurate location at moderate speeds and journal pressures.	H8/f7
	Sliding fit: Where parts are not intended to run freely, but must move and turn freely and locate accurately.	H7/g6
	Locational clearance fit: Provides snug fit for location of stationary parts, but can be freely assembled and disassembled.	H7/h6
Transition	**Locational transition fit:** For accurate location, a compromise between clearance and interference.	H7/k6
	Locational transition fit: For more accurate location where greater interference is permissible.	H7/n6
Interference	**Locational interference fit:** For parts requiring rigidity and alignment with prime accuracy of location but without special bore pressure requirements.	H7/p6
	Medium drive fit: For ordinary steel parts or shrink fits on light sections, the tighter fit usable with cast iron.	H7/s6
	Force fit: Suitable for parts that can be highly stressed or for shrink fits where the heavy pressing force required is impractical.	H7/u6

5.4.1   LOOSE RUNNING FIT (EXAMPLE)

Determine the loose running fit tolerance for a hole and shaft that has a basic diameter of 25.000 mm. From Table 5.4, the specification for a loose running fit is 25H11/25c11, as detailed in Table 5.5.

TABLE 5.5
Hole and Shaft Sizes for Loose Running Fit (H11/c11)

	Hole	Shaft
Tolerance grade	0.130 mm	0.130 mm
Upper deviation	+0.130 mm	−0.110 mm
Lower deviation	0	−0.240 mm
Maximum diameter	25.130 mm	24.890 mm
Minimum diameter	25.000 mm	24.760 mm
Average diameter	25.065 mm	24.825 mm

Maximum clearance $C_{max} = D_{max} - d_{min} = (25.130 - 24.760) = 0.370$ mm
Minimum clearance $C_{min} = D_{min} - d_{max} = (25.000 - 24.890) = 0.110$ mm

There are two forms of showing the tolerances on a drawing:

The first method:

Hole	Shaft
Ø25.000 mm + 0.130/–0 mm	Ø25.000 mm – 0.110/–0.240 mm

The second method:

Hole	Shaft
Ø25.000 mm	Ø24.825 mm
Ø25.130 mm	Ø24.760 mm

This second method is preferred from a machining point of view. It gives the machinist a specific dimension to aim for; if the mean dimension is calculated and if he or she misses it there is still a small margin to recover the machined feature and still remain within the tolerance zone. This method is sometimes referred to as maximum metal conditions.

5.4.2 LOCATION CLEARANCE FIT (EXAMPLE)

Determine the location clearance fit tolerance for a hole and shaft that has a basic diameter of 25.000 mm.

From Table 5.4, the specification for a location clearance fit is 25H7/25h6, as detailed in Table 5.6.

TABLE 5.6
Hole and Shaft Sizes for Location Clearance Fit (H7/h6)

	Hole	Shaft
Tolerance grade	0.021 mm	0.013 mm
Upper deviation	0.021 mm	0.000 mm
Lower deviation	0.000 mm	–0.013 mm
Maximum diameter	25.021 mm	25.000 mm
Minimum diameter	25.000 mm	24.987 mm
Average diameter	25.011 mm	24.994 mm
Maximum clearance	$C_{max} = D_{max} - d_{min} = (25.021 - 24.987) = 0.034$ mm	
Minimum clearance	$C_{min} = D_{min} - d_{max} = (25.000 - 25.000) = 0.000$ mm	

As with the example in Section 5.4.1, there are two methods for showing the tolerances.

The first method:

Hole	Shaft
Ø25.000 mm + 0.021/–0 mm	Ø25.000 mm + 0/+0.013 mm

The second method:

Hole	Shaft
Ø25.000 mm	Ø24.987 mm
Ø25.021 mm	Ø25.000 mm

5.5   SURFACE FINISH

Any discussion about limits and fits invariably brings up the question of surface finish. It is impractical to apply a close tolerance such as 25g6 (0.013) to a machined part that has a surface roughness in the order of, say, 12.5 microns.

Table 5.7 gives an indication of the surface roughness obtained by the various manufacturing processes.

TABLE 5.7

Surface Roughness Values Obtainable by Standard Manufacturing Processes

	Manufacturing Process	Obtainable with Difficulty	Normally Obtainable	Roughing
Casting	Sand casting	0.8–1.6	6.3–12.5	
	Permanent mold		1.6–6.3	12.5–25
	Die casting		0.8–3.2	
Manual	Forging	1.6–3.2	3.2–25	
	Extrusion	0.4–0.8	0.8–6.3	
	Rolling	0.4–0.8	0.8–3.2	
Surface process	Flame cut		25–50	
	Hack saw		6.3–50	
	Bandsaw, chipping		3.2–50	
	Filing	0.8–1.6	1.6–12.5	
	Emery polish	0.1–0.4	0.4–1.6	1.6–3.2
Machining	Shell milling	1.6–3.2	3.2–25	25–50
	Drilling	3.2–6.3	6.3–25	
	Planing and shaping		1.6–12.5	
	Face milling	0.8–1.6	1.6–12.5	12.5–50
	Turning	0.2–1.6	1.6–6.3	6.3–50
	Boring	0.2–1.6	1.6–6.3	6.3–50
	Reaming	0.4–0.8	0.8–6.3	6.3–12.5
	Cylindrical grinding	0.025–0.4	0.4–3.2	3.2–6.3
	Centerless grinding	0.05–0.4	0.4–3.2	
	Internal grinding	0.024–0.4	0.4–3.2	3.2–6.3
	Surface grinding	0.025–0.4	0.4–3.2	3.2–6.3
	Broaching	0.2–0.8	0.8–3.2	3.2–6.3
	Super finishing	0.025–0.1	0.1–0.4	
	Honing	0.025–0.1	0.1–0.4	
	Lapping	0.006–0.05	0.05–0.4	
Gear Manufacturing	Milling, spiral bevel	1.6–3.2	3.2–12.5	12.5–25
	Milling, with form cutter	1.6–3.2	3.2–12.5	12.5–50
	Hobbing	0.8–3.2	3.2–12.5	12.5–50
	Shaping	0.4–1.6	1.6–12.5	12.5–250
	Planing	0.4–1.6	1.6–18.5	12.5–50
	Shaving	0.4–0.8	0.8–3.2	
	Grinding (crisscross)	0.4–0.8	0.8–1.6	
	Grinding	0.1–0.4	0.4–0.8	
	Lapping	0.05–0.2	0.2–0.8	

6 Thick Cylinders

6.1 INTRODUCTION

When the wall thickness of a pressure vessel such as a cylinder or pipe is greater than one-tenth of the inside radius of the vessel, the meridional and circumferential (hoop) stresses cannot be considered uniform throughout the wall thickness and the radial stress will not be negligible. At this point the vessel will be considered to be a thick-walled cylinder as in Figure 6.1.

$$\left(\sigma_r + d\sigma_r\right)\left(r + dr\right)d\theta = \sigma_r r d\theta + 2\sigma_c dr \frac{dr}{2}$$

i.e.
$$r d\theta_r + \sigma_r dr = \sigma_c dr$$

or
$$\sigma_r + r\frac{d\sigma_r}{dr} = \sigma_c \tag{6.1}$$

If the longitudinal stress and strain are denoted by σ_l and ε_l, respectively then:

$$\varepsilon_l = \frac{\sigma_l}{E} - \nu\left(\frac{\sigma_r + \sigma_c}{E}\right)$$

It is assumed that ε_l is constant across the thickness, i.e., a plane cross section of the cylinder remains plane after the application of pressure, and that σ_l is also of uniform thickness, both assumptions being reasonable on planes that are remote from the ends of the cylinder.

6.2 A THICK-WALLED CYLINDER SUBJECT TO INTERNAL AND EXTERNAL PRESSURES

Consider a thick-walled cylinder as shown in Figure 6.1(a) with internal and external radii r_1 and r_2, respectively, and subjected to internal and external pressures p_1 and p_2, respectively. Figure 6.1(b) shows the stresses acting on an element of radius r and thickness dr, subtending an angle $d\theta$ at the center.

The radial and circumferential stresses, σ_r and σ_c, are assumed to be compressive, which in this context is considered positive.

Tensile stresses are positive; compressive stresses are negative.

The pressures p, p_1, p_2, and p_f are negative.

The following relationships are assumed for the strains ε_1, ε_2, and ε_3 associated with the stresses σ_1, σ_2, and σ_3. See Figure 6.2 for the orthogonal stress relationship.

$$\varepsilon_1 = \sigma_1/E - \nu\sigma_2/E - \nu\sigma_3/E$$
$$\varepsilon_2 = \sigma_2/E - \nu\sigma_1/E - \nu\sigma_3/E$$
$$\varepsilon_3 = \sigma_3/E - \nu\sigma_1/E - \nu\sigma_2/E$$

It follows from these assumptions that $\sigma_r + \sigma_c$ is a constant; this will be denoted by 2a.

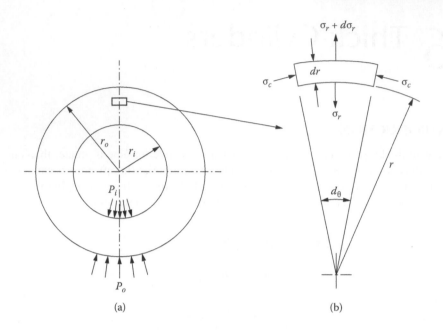

FIGURE 6.1   Elements of a thick-walled cylinder.

TABLE 6.1

Nomenclature

p_1 = Internal pressure (pa)	r = Radius at point of analysis (m)
p_2 = External pressure (Pa)	r_1 = Inside radius of cylinder (m)
p_f = Interface pressure (Pa)	r_2 = Outside radius of cylinder (m)
σ_r = Radial stress – compressive (Pa)	r_f = Interface radius of cylinder (m)
σ_c = Circumferential stress – compressive (Pa)	ε_r = Radial strain
σ_a = Axial/longitudinal stress – tensile (Pa)	ε_c = Circumferential strain
E = Modulus of elasticity (Pa)	ε_a = Axial/longitudinal strain
v = Poisson's ratio	u = Radial deflection (m)
v_h = Poisson's ratio – hub	u_s = Radial deflection of shaft (m)
v_s = Poisson's ratio – shaft	u_h = Radial deflection of hub (m)
d = Diameter at point or analysis (m)	u_t = Radial deflection of hub and shaft (m)
d_1 = Inside diameter of cylinder (m)	δr_h = Radial increase in hole (m)
d_2 = Outside diameter of cylinder (m)	δr_s = Radial decrease in hole (m)

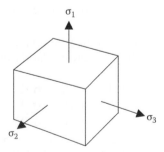

FIGURE 6.2   Stress relationships.

Thus: $\sigma_c = 2a - \sigma_r$ (6.2)

Substituting Equation (6.2) into Equation (6.1):

$$\sigma_r + r\frac{d\sigma_r}{dr} = 2a - \sigma_r$$

or $$2\sigma_r r + r^2 \frac{d\sigma_r}{dr} - 2ar = 0$$

Multiplying both sides by r

i.e., $$\frac{d}{dr}\left(\sigma_r r^2 - ar^2\right) = 0$$

Therefore $$\sigma_r r^2 - ar^2 = b'$$

or $$\sigma_r = a + \frac{b}{r^2}$$ (6.3)

and from Equation (6.2) $$\sigma_c = a - \frac{b}{r^2}$$ (6.4)

Equations (6.3) and (6.4) are known as *Lamé's equations*; in any given application there will always be two conditions sufficient to solve for the constants a and b, together with the radial and circumferential stresses at any radius r, which can then be evaluated.

6.3   GENERAL EQUATIONS FOR A THICK-WALLED CYLINDER SUBJECT TO AN INTERNAL PRESSURE

Considering the common case of a cylinder subjected to an internal pressure only, see Figure 6.3.

$$\sigma_r = P_i \quad \text{when } r = r_i$$

and $$\sigma_r = 0 \quad \text{when } r = r_o$$

$$P_i = a + \frac{b}{r_i^2}$$

and $$0 = a + \frac{b}{r_o^2}$$

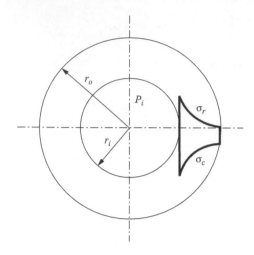

FIGURE 6.3 Thick cylinder subject to an internal pressure.

from which

$$a = -P_i \frac{r_i^2}{r_o^2 - r_i^2}$$

and

$$b = P_i \frac{r_o^2 r_i^2}{r_o^2 - r_i^2}$$

Therefore

$$\sigma_r = a + \frac{b}{r^2}$$

$$= -P_i \frac{r_i^2}{r_o^2 - r_i^2}\left(1 + \frac{r_o^2}{r^2}\right)$$

and

$$\sigma_c = a - \frac{b}{r^2}$$

$$= -P_i \frac{r_i^2}{r_o^2 - r_i^2}\left(1 + \frac{1}{r^2}\right)$$

The maximum radial and circumferential stresses occur at $r = r_i$, when $\sigma_r = P$.

$$\sigma_c = -P_i \frac{r_o^2 + r_i^2}{r_o^2 - r_i^2} \quad \text{the negative sign indicates tension.} \tag{6.5}$$

6.4   THE GENERAL EQUATION FOR A THICK-WALLED CYLINDER SUBJECT TO INTERNAL AND EXTERNAL PRESSURES

The general equation for a thick-walled cylinder subject to internal and external pressures can easily be obtained from Equations (6.3) and (6.4).

Consider a cylinder with an internal diameter d_i, subject to an internal pressure P_i. The outside diameter d_o is subject to an external pressure P_o. The radial pressures at the surfaces are the same as the applied pressures; therefore:

$$\sigma_r = a + \frac{b}{r^2}$$

$$\sigma_c = a - \frac{b}{r^2}$$

(6.6)

as the radial pressures at the surfaces are the same as the applied pressures; therefore

$$-P_i = a + \frac{b}{r_i^2}$$

$$-P_o = a + \frac{b}{r_o^2}$$

The general equation for a thick-walled cylinder is

$$\text{Subtracting } b = \frac{\left[(P_o - P_i) r_i^2 r_o^2 \right]}{\left(r_o^2 - r_i^2 \right)}$$

$$\text{Substituting this equation } a = \frac{\left(P_i \cdot r_i^2 - P_o \cdot r_o^2 \right)}{\left(r_o^2 - r_i^2 \right)}$$

and the resulting general equation is shown as follows:

$$\sigma_r = \frac{\left(P_i \cdot r_i^2 - P_o \cdot r_o^2 \right)}{\left(r_o^2 - r_i^2 \right)} + \frac{(P_o - P_i) \cdot r_i^2 \cdot r_o^2}{r^2 \cdot \left(r_o^2 - r_i^2 \right)}$$

$$\sigma_c = \frac{\left(P_i \cdot r_i^2 - P_o \cdot r_o^2 \right)}{\left(r_o^2 - r_i^2 \right)} - \frac{(P_o - P_i) \cdot r_i^2 \cdot r_o^2}{r^2 \cdot \left(r_o^2 - r_i^2 \right)}$$

If the external pressure is zero, the equation reduces to

$$\sigma_r = \frac{r_i^2 \left(r^2 - r_o^2 \right)}{r^2 \cdot \left(r_o^2 - r_i^2 \right)} P_i \quad \text{and} \quad \sigma_c = \frac{r_i^2 \cdot \left(r^2 + r_o^2 \right)}{r^2 \cdot \left(r_o^2 - r_i^2 \right)} P_i$$

If the internal pressure is zero, the equation reduces to:

$$\sigma_r = \frac{r_o^2 \left(r_i^2 - r^2 \right)}{r^2 \cdot \left(r_o^2 - r_i^2 \right)} P_o \quad \text{and} \quad \sigma_c = -\left(\frac{r_o^2 \cdot \left(r_i^2 + r^2 \right)}{r^2 \cdot \left(r_o^2 - r_i^2 \right)} P_o \right)$$

If the cylinder has closed ends, the axial stress can be found using the force equilibrium considerations only.

The pressure P_i acts on an area given by πr_i^2.
The pressure P_o acts on an area given by πr_o^2.
The axial stress σ_z acts on an area given by $\pi(r_o^2 - r_i^2)$.

Force equilibrium then gives

$$\sigma_z = \left(\frac{P_i \cdot r_i^2 - P_o \cdot r_o^2}{r_o^2 - r_i^2} \right)$$

EXAMPLE 6.1: INTERFERENCE FIT

When considering a force fit of a shaft into a hub, generally they are manufactured from the same material. This is a simple example of a compound cylinder.

Consider a press fit of a shaft into a hub. The compression of the shaft and the expansion of the hub result in a compressive pressure at the interface between the shaft and hub. The conditions are shown in Figure 6.4 and the individual components are shown in Figure 6.5.

The radial interference δr_f = sum of the shaft deflection δr_s and the hole deflection δr_h.

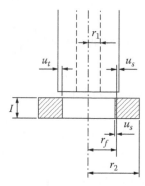

FIGURE 6.4   Interference fit.

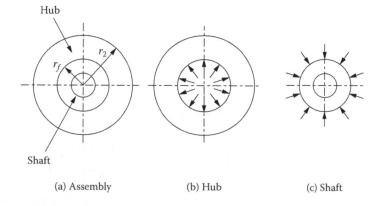

FIGURE 6.5   Components of shaft and hub.

The longitudinal pressure, and hence σ_a, is assumed to be zero, and the internal pressure in the shaft hole and the external pressure outside the hub are also assumed to be zero.

$$E\varepsilon_c = E\frac{u}{r} = \sigma_c - \upsilon\sigma_a - \upsilon\sigma_r = \sigma_c - \upsilon\sigma_r \qquad (6.7)$$

The radial increase in the hole diameter = u_h.

$$u_h = \left(\frac{r_f}{E_h}\right)(\sigma_c - \upsilon\sigma \cdot \sigma_r)$$

The radial decrease in the shaft diameter = u_s

$$u_s = \left(\frac{r_f}{E_s}\right)(\sigma_c - \upsilon\sigma \cdot \sigma_r)$$

The total interference $u_t = u_h + u_s$

The equations for the hole:

$$\sigma_r = \frac{r_f^2 \cdot \left(r_f^2 - r_2^2\right)}{r_f^2 \cdot \left(r_2^2 - r_f^2\right)} P_f$$

$$\sigma_c = \frac{\left(r_2^2 + r_f^2\right)}{r_2^2 - r_f^2} P_f$$

The equations for the shaft:

$$\sigma_r = \frac{r_f^2 \cdot \left(r_1^2 - r_f^2\right)}{r^2 \cdot \left(r_f^2 - r_1^2\right)} P_f$$

$$\sigma_c = -\left(\frac{\left(r_f^2 + r_1^2\right)}{r_f^2 - r_1^2} P_f\right)$$

The interference equations are:

For the hole:

$$u_h = \frac{r_f \cdot P_f}{E_h}\left(\frac{r_2^2 + r_f^2}{r_2^2 - r_f^2} + \upsilon_h\right)$$

For the shaft:

$$u_s = -\frac{r_f \cdot P_f}{E_s}\left(\frac{r_f^2 + r_1^2}{r_f^2 - r_1^2} - \upsilon_h\right)$$

The total interference is therefore:

$$u_t = u_h + (-u_s)$$

$$= \frac{r_f \cdot P_f}{E_h}\left(\frac{r_2^2 + r_f^2}{r_2^2 - r_f^2} + \upsilon_h\right) + \frac{r_f \cdot P_f}{E_s}\left(\frac{r_f^2 + r_1^2}{r_f^2 - r_1^2} - \upsilon_h\right)$$

If the hub and shaft are manufactured from the same material having an identical modulus of elasticity (E) and Poisson's ratio (v), the above equation will simplify to:

$$u_t = \frac{r_f \cdot P_f}{E}\left(\frac{2r_f^2\left(r_2^2 - r_1^2\right)}{\left(r_2^2 - r_f^2\right)\left(r_f^2 - r_1^2\right)}\right)$$

In a normal engineering application where the shaft is solid, i.e, when $r_1 = 0$, the equation is further simplified to:

$$u_t = \frac{2r_2^2 \cdot r_f \cdot P_f}{E\left(r_2^2 - r_f^2\right)}$$

There is often a requirement to calculate the interface pressure when the radial interference u_t is known (i.e., the shaft interference/2). When calculating the torque that can be transmitted or the force required to either make or separate the interference joint,

$$P_f = \frac{E\left(r_2^2 - r_f^2\right) \cdot u_t}{2 \cdot r_2^2 \cdot r_f}$$

By way of an example, consider a steel shaft 50 mm diameter pressed into a hub that has an outside diameter of 150 mm. The length of the hub is 50 mm and the interference is 0.05 mm. The assumed coefficient of friction is $\mu = 0.15$.

The hub and shaft are both steel with the modulus of elasticity (E) = 210×10^9 Pa and the Poisson's ratio $\upsilon = 0.3$.

$$P_f = \frac{E\left(r_2^2 - r_f^2\right) \cdot u_t}{2 \cdot r_2^2 \cdot r_f}$$

$$= \frac{210 \times 10^9 \cdot \left(0.05^2 - 0.025^2\right) \cdot 0.5 \times 10^{-6}}{2 \cdot 0.15^2 \cdot 0.025}$$

$$= 6.013 \times 10^6 \text{ Nm}$$

Torque:

$$T = \pi \cdot d \cdot L \cdot P_f \cdot \mu \cdot r_f$$

$$= \pi \times 50 \text{ mm} \times 50 \text{ mm} \times 204.167 \times 10^6 \text{Pa} \times 0.15 \times 25 \text{ mm}$$

$$= 6.013 \times 10^3 \text{ Nm}$$

EXAMPLE 6.2: RADIAL DISTRIBUTION OF STRESS

A steel cylinder has an outside diameter of 200 mm and an inside diameter of 100 mm. The cylinder is subject to an internal pressure of 150 MPa. Determine the radial and circumferential stress distributions and show the results in the form of a spreadsheet. Assume the cylinder has closed ends.

The section is divided into incremental radii increasing from the inner radius of 50.0 mm at 2.5 mm intervals to the outside radius of 100.0 mm. The resultant stress is calculated at each incremental radius. The resultant stresses are tabulated in Table 6.2 and Figure 6.6 shows the results of the analysis.

TABLE 6.2
Spreadsheet Results for Example 6.2

Position	P_i MPa	P_o MPa	r_1 mm	r_0 mm	r mm	Radial Stress MPa	Circ Stress MPa	Axial Stress MPa
1	150	0	50	100	50.0	150.000	250.000	50.000
2	150	0	50	100	52.5	131.406	231.406	50.000
3	150	0	50	100	55.0	115.289	215.289	50.000
4	150	0	50	100	57.5	101.229	201.229	50.000
5	150	0	50	100	60.0	88.889	188.889	50.000
6	150	0	50	100	62.5	78.000	178.000	50.000
7	150	0	50	100	65.0	68.343	168.343	50.000
8	150	0	50	100	67.5	59.739	159.739	50.000
9	150	0	50	100	70.0	52.041	152.041	50.000
10	150	0	50	100	72.5	45.125	145.125	50.000
11	150	0	50	100	75.0	38.889	138.889	50.000
12	150	0	50	100	77.5	33.247	133.247	50.000
13	150	0	50	100	80.0	28.125	128.125	50.000
14	150	0	50	100	82.5	23.462	123.462	50.000
15	150	0	50	100	85.0	19.204	119.204	50.000
16	150	0	50	100	87.5	15.306	115.306	50.000
17	150	0	50	100	90.0	11.728	111.728	50.000
18	150	0	50	100	92.5	8.437	108.437	50.000
19	150	0	50	100	95.0	5.402	105.402	50.000
20	150	0	50	100	97.5	2.597	102.597	50.000
21	150	0	50	100	100.0	0.000	100.000	50.000

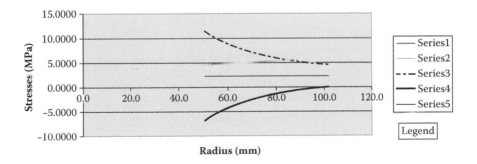

FIGURE 6.6 Thick cylinder stresses.

7 Compound Cylinders

7.1 INTRODUCTION

A compound cylinder consists of two concentric cylinders, as shown in Figure 7.1, where the outer cylinder is shrunk onto an inner cylinder such that the latter is initially in compression prior to the application of any internal pressure.

The final stresses in the assembly will then be the result of those due to prestressing and those due to the internal pressure.

Compound cylinders are used to increase the pressure that can be contained in cylinders. Applications involving compound cylinders include gun barrels, etc.

If the radius of the common surface (or interface diameter) is r_o and the pressure at this surface before the application of any internal pressure is p_o, the initial stresses are determined by considering the two cylinders separately, the boundary conditions for the outer cylinder being $\sigma_r = P_o$ when $r = r_o$ and $\sigma_r = 0$ when $r = r_1$. For the inner cylinder $\sigma_r = P_o$ when $r = r_o$ and $\sigma_r = 0$ when $r = r_2$.

The stresses due to the internal pressure are obtained by considering the cylinder to be homogenous, with $\sigma_r = P$ at $r = r_2$ and $\sigma_r = 0$ at $r = r_1$.

The various stresses are then combined algebraically, as shown in Figure 7.2(a), from which it is evident that the maximum resultant circumferential stress is less than for a homogeneous cylinder of the same cross section with the same internal pressure. Alternatively, for the same maximum stress, a thinner cylinder can be used if it is prestressed, the optimum conditions being when the resultant circumferential stress is the same at the inner surface of each cylinder as shown in Figure 7.2(b).

7.2 SHRINKAGE ALLOWANCE

It is necessary for the inside diameter of the outer cylinder to be slightly smaller than the outside diameter of the inner cylinder to produce a desired initial pressure p_o at the common surface of the compound cylinder. It is common practice for the outer cylinder to be heated until it is possible to slide over the inner cylinder, and when cooled to the ambient temperature, it will generate the required pressure at the common surface. This procedure is termed the *shrinkage allowance*.

Let the circumferential stress at the outer surface of the inner cylinder due to p_o be σ_c'.

The circumferential strain at the outer surface:
$$= \frac{\sigma_c'}{E} = v\frac{p_o}{E}$$

Therefore:
$$= \frac{2r_o}{E}\left(\sigma_c' - vp_n\right)$$

For the outer cylinder, let the circumferential stress at the inner surface due to p_o be σ_c''.

The circumferential strain at the inner surface:
$$= \frac{\sigma_c''}{E} - v\frac{p_o}{E}$$

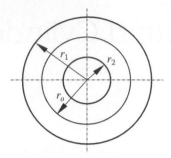

FIGURE 7.1 Compound cylinder.

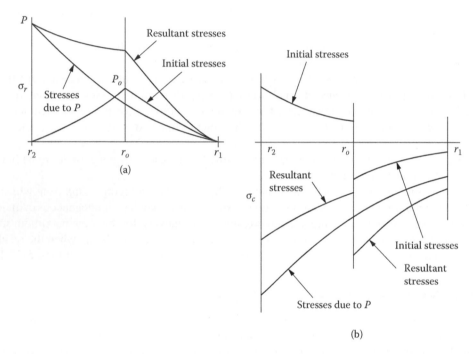

FIGURE 7.2 Algebraic combination of stresses. (a) Stress distribution in a homogeneous cylinder. (b) Stress distribution in a compound cylinder.

Therefore the decrease in diameter $= \dfrac{2r_o}{E}\left(\sigma_c'' - vp_o\right)$

$$\text{Initial difference in diameter} = \frac{2r_o}{E}\left\{\left(\sigma_c' - vp_o\right) - \left(\sigma_c'' - vp_o\right)\right\}$$

$$= \frac{2r}{E}\left(\sigma_c' - \sigma_c''\right)$$

When the materials of the two cylinders have different modulii of elasticity, E_1 and E_2, together with the Poisson's ratio v_1 and v_2, respectively, for the inner and outer cylinders,

$$\text{Shrinkage allowance} = 2r_o\left\{\frac{\sigma_c' - v_1 p_o}{E_1} - \frac{\sigma_c'' - v_2 p_o}{E_2}\right\}$$

EXAMPLE 7.1

Consider a compound cylinder subjected to an internal pressure of 150 MPa.

The radii of the inner, interface, and outer surfaces are, respectively, 100, 140, and 200 mm. Determine the shrinkage pressure necessary for the resultant maximum circumferential stress to be as small as possible and calculate this stress.

Initial stresses: $r_1 = 200$ mm (outside radius)

Let the shrinkage pressure = p $r_o = 140$ mm (interface radius)

                                              $r_2 = 100$ mm (inside radius)

For the inncer cylinder: $p_2 = 150$ MPa

$\sigma_r = p$ at r = 140 mm $p = a_1 + \dfrac{b_1}{140^2}$

                                         $= a_1 + \dfrac{b_1}{19.60 \times 10^3}$

                          $\therefore \quad p = a_1 + 51.02 \times 10^{-6} b_1$

                                         $0 = a_1 + \dfrac{b_1}{100^2}$

                                         $= a_1 + \dfrac{b_1}{10 \times 10^3}$

$\therefore \quad 0 = a_1 + 100 \times 10^{-6} b_1$ (7.1)

$\quad\quad p = a_1 + 51.02 \times 10^{-6} b_1$ (7.2)

Subtracting Equation (7.2) from (7.1) $p = -48.98 \times 10^{-6} b_1$

Therefore: $b_1 = -20.4167 \times 10^3 p$ (7.3)

Substituting Equaion (7.3) into (7.2) $p = a_1 + 51.02 \times 10^{-6} \times (-20.4167 \times 10^3) p$

                                         $p = a_1 - 1.0417 p$

                          $\therefore \quad a_1 = 2.0417 p$ (7.4)

Therefore at the inner surface: $\sigma_c = a_1 - \dfrac{b_1}{100^2}$

                                    $\sigma_c = 2.0417\, p - \dfrac{(2.0417\ p)}{100^2}$

                                    $\sigma_c = 4.0833\, p$

For the outer cylinder:

$\sigma_r = p$ at r = 140 mm $\therefore \ p = a_2 + \dfrac{b_2}{140^2}$

$$= a_2 + 51.02 \times 10^{-6} \ b_2 \qquad (7.5)$$

$\sigma_r = 0$ at r = 200 mm $0 = a_2 + \dfrac{b_2}{200^2}$

$$0 = a_2 + 25.0 \times 10^{-6} \ b_2 \qquad (7.6)$$

Subtracting Equation (7.6) from (7.5): $p = a_2 + 51.02 \times 10^{-6} \ b_2$

$$0 = a_2 + 25.00 \times 10^{-6} \ b_2$$

$$p = \ \ \ + 26.02 \times 10^{-6} \ b_2$$

and $b_2 = 38.432 \times 10^3 \ p \qquad (7.7)$

Substituting Equation (7.7) into (7.5): $p = a_2 + \left(51.2 \times 10^{-6} \times 38.432 \times 10^3\right) p$

$$p = a_2 + 1.9608 \ p$$

$$a_2 = -0.9608 \ p$$

Therefore at the inner surface: $\sigma_c = -0.9608 \ p - \dfrac{b_2}{140^2}$

$$= -0.9608 \ p - \dfrac{38.432 \times 10^3 \ p}{19.600 \times 10^3}$$

$$\sigma_c = -0.9608 \ p - 1.9608 \ p$$

$$\sigma_c = -2.922 \ p$$

Stresses due to the internal pressure:

$\therefore \ \ \sigma_r = 100$ MPa at r = 100 mm $100 = a + \dfrac{b}{100^2}$

$$= a + 100 \times 10^{-6} \ b \qquad (7.8)$$

$\therefore \ \ \sigma_r = 0$ at r = 200 mm $0 = a + \dfrac{b}{200^2}$

$$= a + 25.00 \times 10^{-6} \ b \qquad (7.9)$$

Subtracting Equation (7.9) from (7.8) $100 = a + 100 \times 10^{-6}\, b$

$$0 = a + 25.00 \times 10^{-6}\, b$$

Hence: $100 = 75.00 \times 10^{-6}\, b$

$$b = 1.333 \times 10^{6}$$

Substituting in Equation (7.8) $100 = a + \left(100 \times 10^{-6} \times 1.333 \times 10^{6}\right)$

$$100 = a + 133.33$$

$$a = -33.33$$

Therefore at r = 100 $\sigma_c = a - \dfrac{b}{100^2}$

$$= -33.33 - \frac{1.333 \times 10^{6}}{100^2}$$

$$= -33.33 - 133.3$$

$$\sigma_c = -166.67 \text{ MPa}$$

and at r = 140 mm

$$\sigma_c = a - \frac{b}{140^2}$$

$$= -33.33 - \frac{1.333 \times 10^{6}}{140^2}$$

$$\sigma_c = -101.34 \text{ MPa}$$

For the resultant maximum circumferential stress to be as small as possible, the resultant stresses at the inner surfaces of each cylinder must be equal, as will be evident from Figure 7.2, i.e., −166.67 + 4.0833 p = −101.34 − 2.922 p.

From which: $-166.67 + 101.34 = 2.922\,p - 4.0833\,p$

$$-65.33 = 7.0053\,p$$

$$\frac{66.67}{-7.0053} = p$$

$$9.326 \text{ MPa} = p$$

Resultant maximum stress: $-101.34 - (3 \times 9.326) = 129.32 \text{ MPa}$

EXAMPLE 7.2

A compound steel cylinder has a bore of 75.0 mm with an external diameter of 175.0 mm; the diameter of the interface is 120.0 mm. Determine the radial pressure at the interface, which must be provided by shrinkage. If the resultant maximum circumferential stress in the inner cylinder under a superimposed internal pressure of 60 MPa is to be half the value of the maximum hoop tension, which would be produced in the inner cylinder if that cylinder alone were subjected to an internal pressure of 60 MPa.

Determine the final hoop tensions at the inner and outer surfaces of both cylinders under the internal pressure of 60 MPa and sketch a graph to show the circumferential stress varies across the cylinder wall.

1. Initial stresses in the inner cylinder.

Let the shrinkage pressure be p. $r_1 = 87.5$ mm (outside radius)

$r_0 = 60.0$ mm (interface radius)

$r_2 = 37.5$ mm (inside radius)

$P = 60.0$ MPa

$\sigma_r = p$ at r = 60.0 mm $p = a_1 + \dfrac{b_1}{60.0^2}$

hence: $p = a_1 + 277.778 \times 10^{-6} \, b_1$ (7.10)

and $\sigma_r = 0$ at r = 37.5 mm $0 = a_1 + \dfrac{b_1}{37.5^2}$

$0 = a_1 + 711.111 \times 10^{-6} \, b_1$ (7.11)

Subtracting Equation (7.10) from (7.11) $p = a_1 + 277.778 \times 10^{-6} \, b_1$

$p = a_1 + 711.111 \times 10^{-6} \, b_1$

$p = -433.33 \times 10^{-6} \, b_1$

$b_1 = -2307.705 \, p$

Substituting in Equation (7.10) $p = a_1 + \left(277.78 \times 10^{-6} \times 2307.705 \, p \right)$

$p = a_1 - 0.64103 \, p$

$1.641 p = a_1$

Therefore at the inner surface: $\sigma_c = a_1 - \dfrac{b_1}{37.5^2}$

$= 1.641 p + \dfrac{2307.705 \, p}{37.5^2}$

$= 1.641 p + 1.641 p$

$\sigma_c = 3.282 \, p$

2. Stresses due to the internal pressure.

$\sigma_r = 60$ MPa at r = 37.5 mm $\qquad 60 = a_2 + \dfrac{b_2}{37.5^2}$

$$60 = a_2 + 711.111 \times 10^{-6}\, b_2 \qquad (7.12)$$

and at $\sigma_r = 0$ at r = 87.5 mm $\qquad 0 = a_2 + \dfrac{b_2}{87.5^2}$

$$0 = a_2 + 130.612 \times 10^{-6}\, b_2 \qquad (7.13)$$

Subtracting Equation (7.13) from (7.12)

$$60 = a_2 + 711.111 \times 10^{-6}\, b_2$$
$$0 = a_2 + 130.612 \times 10^{-6}\, b_2$$
$$60 = 580.50 \times 10^{-6}\, b_2$$
$$103.359 \times 10^{3} = b_2$$

Substituting in Equation (7.12)

$$60 = a_2 + \left(711.111 \times 10^{-6} \times 103.359 \times 10^{3}\right)$$
$$60 = a_2 + 73.50$$
$$60 - 73.50 = a_2$$
$$-13.50 - a_2$$

Therefore, at the inner surface: $\qquad \sigma_c = a_2 - \dfrac{b_2}{37.5^2}$

$$\sigma_c = a_2 - 711.111 \times 10^{-6}\, b_2$$
$$\sigma_c = -13.50 - \left(711.111 \times 10^{-6} \times 103.59 \times 10^{3}\right)$$
$$= -13.50 - 73.50$$
$$= -87.0 + 2.8928\, p$$

3. Inner cylinder alone with internal pressure.

$\sigma_r = 60$ MPa at r = 37.5 mm $\qquad 60 = a_3 + \dfrac{b_3}{37.5^2}$

$$60 = a_3 + 711.111 \times 10^{-6} \, b_3 \qquad\qquad (7.14)$$

$\sigma_r = 0$ at r = 60.0 mm $\qquad\qquad 0 = a_3 + \dfrac{b_3}{60.0^2} \qquad\qquad (7.15)$

Subtracting Equation (7.15) from (7.14)

$$60 = a_3 + 711.111 \times 10^{-6} \, b_3$$
$$0 = a_3 + 277.78 \times 10^{-6} \, b_3$$
$$60 = \quad + 433.333 \times 10^{-6} \, b_3$$
$$138.462 \times 10^3 = b_3$$

Substituting in Equation (7.14)

$$60 = a_3 + (711.111 \times 10^{-6} \times 138.462 \times 10^3)$$
$$a_3 = -38.4616$$

At the inner surface:

$$\sigma_c = a_3 - \dfrac{b_3}{37.5^2}$$
$$\sigma_c = -38.4616 - \dfrac{138.462 \times 10^3}{37.52}$$
$$\sigma_c = -136.923 \, \text{MPa}$$
$$-87.0 + 2.8928 \, p = \dfrac{-136.923}{2}$$
$$p = 6.4085 \, \text{MPa}$$

4. Initial stresses in the outer cylinder:

$\sigma_r = p$ at $r = 60.0$ mm $\therefore\ p = a_4 + \dfrac{b_4}{60^2}$

$$p = a_4 + 277.78 \times 10^{-6}\,b_4 \qquad\qquad (7.16)$$

$\sigma_r = 0$ at $r = 87.5$ mm $\therefore\ 0 = a_4 + \dfrac{b_4}{87.5^2}$

$$0 = a_4 + 130.612 \times 10^{-6}\,b_4 \qquad\qquad (7.17)$$

Subtracting Equation (7.17) from (7.16)

$$p = a_4 + 277.780 \times 10^{-6}\,b_4$$
$$0 = a_4 + 130.612 \times 10^{-6}\,b_4$$
$$p = \quad + 147.166 \times 10^{-6}\,b_4$$
$$6795.06\,p = b_4$$

Substituting in Equation (7.16)

$$p = a_4 + (277.78 \times 10^{-6} \times 6795.06\,p)$$
$$p = a_4 + 1.8875\,p$$
$$-0.8875\,p = a_4$$

The resultant circumferential stresses in the compound cylinder are then as follows:

Inner cylinder, inner surface:

$$\sigma_{c1} = \frac{-136.923}{2}\,\text{MPa}$$
$$\sigma_{c1} = -68.4615\,\text{MPa}$$

Inner cylinder, outer surface:

$$\sigma_{c2} = \left(a_1 - \frac{b_1}{60.0^2}\right) + \left(a_2 - \frac{b_2}{60.0^2}\right)$$
$$= \left(1.6410\,p + \frac{-2307.705\,p}{60.0^2}\right) + \left(-13.50 - \frac{103.36 \times 10^3}{60.0^2}\right)$$
$$= 1.00 \times 6.4085 - 42.211$$
$$\sigma_{c2} = -35.8025\,\text{MPa}$$

Outer cylinder, inner surface:

$$\sigma_{c3} = \left(a_4 - \frac{b_4}{60.0^2}\right) + \left(a_2 - \frac{b_2}{60.0^2}\right)$$

$$\sigma_{c3} = \left(-0.8875\,p - \frac{6795.06\,p}{60.0^2}\right) + \left(-13.50 - \frac{103.359}{60.0^2}\right)$$

$$\sigma_{c3} = -59.995\ \text{MPa}$$

Outer cylinder, outer surface:

$$\sigma_{c4} = \left(a_4 - \frac{b_4}{87.5^2}\right) + \left(a_2 - \frac{b_2}{87.5^2}\right)$$

$$\sigma_{c4} = \left(-0.8875\,p - \frac{6795.06\,p}{87.5^2}\right) + \left(-13.50 - \frac{103.359 \times 10^3}{87.5^2}\right)$$

$$\sigma_{c4} = -38.375\ \text{MPa}$$

The variation in the circumferential stress across the cylinder is shown in Figure 7.3.

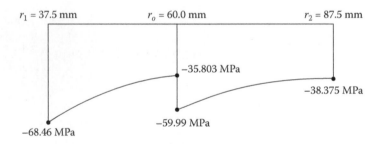

FIGURE 7.3   Variation in the circumferential stress across the cylinder.

FURTHER READING

Warren C. Young. *Roark's Formulas for Stress and Strain*. 6th ed. New York: McGraw-Hill, 1989.

8 The Design and Analysis of Helical Compression Springs Manufactured from Round Wire

8.1 ELASTIC STRESSES AND DEFLECTIONS OF HELICAL COMPRESSION SPRINGS MANUFACTURED FROM ROUND WIRE

8.1.1 INTRODUCTION

This chapter covers the design data for helical compression springs manufactured from round wire and acting within the elastic range of the material. It is in four sections:

8.1: Elastic Stresses and Deflections of Helical Compression Springs Manufactured from Round Wire
8.2: Allowable Stresses for Helical Compression Springs Manufactured from Round Wire
8.3: Notes on the Design of Helical Compression Springs Made from Round Wire
8.4: Nested Helical Compression Springs

The design of a helical compression spring to meet specific requirements will require going through a number of iterations before a suitable spring design is finally established.

The procedure outlined in this chapter will enable a suitable spring design to be arrived at very quickly when all the constraints are clearly identified.

The design process outlined below identifies the use of each subsection in Section 8.1, although not all subsections will be required for any one spring design. This process may have to be iterated a number of times before a suitable design will be established.

1. The type of end coil will need to be decided. These will influence both the stresses and the number of active coils in the spring. See Section 8.3, Notes on the Design of Helical Compression Springs Made from Round Wire.
2. Establish a suitable material and a value for the working stress. Guidance on this is given in Section 8.2, Allowable Stresses for Helical Compression Springs Manufactured from Round Wire.
3. The external diameter (Do) or internal diameter (Di) of the spring is often fixed by the design problem or can be easily established. Section 8.1.5 may be used to determine d and hence c and D.
4. Section 8.1.5 also covers the spring rate, deflection characteristics, and number of turns.
5. The spring will require to be checked that it conforms to the limitations of Section 8.1.3.
6. The closed length characteristic is checked using Section 8.1.7.
7. The buckling characteristic is checked using Section 8.1.8.
8. The outside diameter of the spring will increase slightly when loaded. This increase is generally less than 2%, but may be accurately determined, if necessary, by using Section 8.1.10.
9. Additional stresses may be incurred if the spring is subject to transverse loading (see Section 8.1.9).

10. Helix warping produces a stress increase over and above that which is predicted in Section 8.1.11. The increase is slight (being less than 5% for springs for which $N > 11$) and needs to be considered only in spring designs near their stress limit or where fatigue is the governing failure mode. See Section 8.1.11, for springs with ends closed and ground.

8.1.2 Notation

P	= Axial load applied to spring
q	= Shear stress due to P
δ	= Deflection due to P
S	= Rate or stiffness of spring
G	= Modulus of rigidity
D	= Mean diameter of coil
d	= Diameter of wire
c	= D/d spring index
n	= Number of active coils
N	= Total number of coils
L_o	= Free length
L_c	= Solid length including end coils
$L_{c'}$	= Solid length omitting end coils (nd)
D_o	= Outside diameter of coil
D_i	= Inside diameter of end coils
C_1 to C_5	= Functions of c

8.1.3 Notes

The data in this chapter are subject to the following limitations.

1. The helix angle of a spring should not exceed 10°; therefore, for an unloaded compression spring,

$$L_o < 0.55 n_o D + d' \tag{8.1}$$

where values of d' are given in Table 8.1.
2. The deflection per coil under load should not be more than about a quarter of the coil diameter, D.

Most springs will conform to conditions 1 and 2, but if these limitations are exceeded, then significant errors will arise; in this case it should be considered that the single spring should be replaced by a nest of two or more springs, each conforming to the conditions above (see Section 8.4).

TABLE 8.1
Allowances for End Coil Forms

End coil formation	d'
One coil at each end closed and ground flat	1.5d
End coils unclosed but ground	−0.5d
End coils unclosed and unground	d
One coil at each end closed but unground	3.0d

TABLE 8.2
Material Specifications

Spring Material	Specification
Steel	S201 or S202
Stainless steel	S205
Nimonic 90	HR501HD
Nimonic 90	HR502
Phosphor bronze	BS 384

8.1.4 COMPRESSION SPRING CHARACTERISTICS

8.1.4.1 Material Specifications

The materials covered in this chapter (but not limited to) are as shown in Table 8.2.

8.1.4.2 Wire Diameter

The wire diameter may be quoted in either imperial or metric diameters using readily available sizes.

Tables 8.7 and 8.8 gives approximately allowable stresses for standard wire sizes between 0.0076 and 0.300 in. and 0.250 and 12.00 mm.

8.1.4.3 Mean Diameter

The mean diameter is a basic characteristic of the helical spring. Although the outside or inside diameter of the spring may be quoted, the mean diameter will be found by either subtracting or adding the wire diameter to these diameters.

8.1.4.4 Spring Index c

The spring index is found by dividing the mean diameter D by the wire diameter d.

The preferred index is between 4 and 12. If the index is less than 4 the spring may be too difficult to manufacture and be too highly stressed; if the index is greater than 12 the spring may be too flimsy, tangle easily, or the coils may slip over each other as the spring is compressed to solid.

The larger the index the greater the deflection in comparison to the solid length.

8.1.4.5 Spring Rate

This is another characteristic of the helical spring: in a helical spring design the deflection is proportional to the force acting on the spring. This is the same for most springs where the spring rate is linearly proportional.

8.1.4.6 Number of Active Coils

The number of active coils is those coils that actually deflect when the spring is either compressed or extended when subjected to the force.

8.1.4.7 Total Number of Coils

The total number of coils equals the number of active coils plus the number of end coils, if any, and may be taken as follows:

Type of Spring Ends

Closed and ground flat = N = n + 2   where N = total number of coils
Closed but not ground = N = n + 2   n = number of active coils
Open   = N = n

8.1.4.8   Solid Length

This is the length of the spring when all the coils are compressed and touching each other plus the type of spring end.

If the design calls for plating or an alternative coating, bear this in mind as this will increase the solid length.

The solid length should be specified as the maximum dimension. It should not be the same as the calculated nominal solid length.

8.1.4.9   Initial (Free) Length

This is the overall length of the spring in its fully unloaded condition.

If loads are specified then the free length should be a reference dimension. This will allow the spring maker to vary the free length to maintain the spring force relative to the spring deflection or specified length.

8.1.4.10   Clearance at Maximum Load

The free length of the spring should be chosen so that sufficient clearance is left between coils when the spring is compressed by the maximum working load.

This clearance should be not less than 15% of the deflection of the spring from the free length to the solid length.

8.1.4.11   Direction of Wind

A helical spring can be wound either right or left hand. If the direction of wind is not specified, it can be coiled either way.

If the springs are nested then the wind must be in opposite directions to eliminate coil binding.

When the spring is fitted over a screw thread then the coiling should be in the opposite direction of the screw thread.

8.1.4.12   Allowable Stresses

The allowable stresses will be determined by the material being specified and the operating temperature of the spring.

As a rule of thumb it is generally considered that if the maximum working stress in the spring is kept below 40% of the allowable stress for the material specified, then the spring will have an unlimited life. This will lead to a bulky spring, but if weight is not a problem, then this will be acceptable.

Where spring weight is critical care will be required in calculating the maximum stress in the spring for the particular application, and the life of the spring will need to be calculated.

Tables 8.9 to 8.11 tabulate the temperature reduction factors and maximum temperatures required for the standard materials covered in this chapter.

8.1.4.13   Finish

Depending on the operating environment the spring will be working in, it may be desirable to specify a finish to the spring to protect against corrosion, etc. Care will be required in selecting the correct specification.

Some finishes or plating may have a detrimental effect on the fatigue life of the spring that could lead to premature failure.

8.1.5   STATIC SHEAR STRESS

8.1.5.1   Basic Formulas

8.1.5.1.1   Spring Rate

The rate or stiffness of a spring is the load required to produce a unit of deflection.

$$S = \frac{P}{\delta} \quad \text{or} \quad S = \frac{8 \times P \times D}{\pi \times d^3}$$

Note that several authorities introduce a correction factor for variation in spring index, but this factor is small and can be safely ignored. The number of working coils cannot be precisely defined except by careful tests on individual springs, and the possibility of error from this source is greater than the effect of possible correction factors.

8.1.5.2   Maximum Shear Stress

This is the stress at the surface of the spring wire produced by a given load. The stress is higher on the inner surface of the coil.

Uncorrected for coil curvature:

$$q = \frac{8 \times P \times D}{\pi \times d^3}$$

With curvature correction:

$$q = \frac{8 \times P \times D}{\pi \times d^3} \times K \qquad K = \frac{(c + 0.2)}{(c - 1)} \quad \text{(Stress concentration factor for curvature where } c = D/d)$$

8.1.5.2.1   Note A

This curvature correction factor was first proposed by National Physical Laboratory (NPL) in England in Spring Design Memorandum 1, published by the Ministry of Supply in 1943. It gives values very close to those obtained with Wahl's formulas:

$$K = \frac{4c - 1}{4c - 4} + \frac{0.615}{c}$$

The previous formula is much simpler to apply.

8.1.5.2.2   Note B

The correction factor should always be applied to springs subject to fatigue or compression to closure for extended periods of time.

The uncorrected formulas may be applied where fatigue or sustained compression (e.g., clamping, etc.) is not involved.

Alternatively:

$$q = \frac{G \times \delta}{\pi \times c^2 \times L_c}$$

8.1.5.3   Useful Relations

Note that imperial or Système International (SI) units may be used providing that they are consistent.

The problem of design is to find the dimensions of a spring for a specified performance. This is the inverse of the problem of finding rate and stress, when all the dimensions are known. The design problem can be solved from the basic formulas by a process of successive approximation, but the use of the following functions of the spring index gives the solution more readily:

$$D_o = (c+1) \times d \qquad D_i = (c-1) \times d$$

$$d = \frac{D_o}{c+1} \qquad d = \frac{D_i}{c-1} \qquad c = \frac{D_o + D_i}{D_o - D_i} \qquad c = \frac{D_o - d}{d}$$

D_o is used in preference to D, since the problem may be to find the remaining dimensions when the spring housing diameter is given. values for the above functions are given in Tables 8.3 and 8.4 (covering imperial and metric (SI) springs, respectively), for the range of values of c from 2.0 to 12.0. Wherever possible, c should be restricted to the range from 3.0 to 10.0; greater than 10.0 may sometimes be necessary to avoid excessive lengths of springs.

A single spring of circular section is completely defined if any four of the following six parameters are known, P, δ, q, D_o, d, and n. (Note that S, L_{cl}, c, etc., are included in the forgoing since

$$S = \frac{P}{\delta} \qquad L_c = n \times d \qquad c = \frac{D}{d}$$

If only three parameters are specified, a value will need to be assumed for one of the unspecified parameters before the design can be completed. A satisfactory spring can be designed by taking a value of c in the middle of the range.

8.1.5.4 Relationships

Values of design functions, C_1, C_2, C_3, C_4, and C_5.

$$C_1 = \sqrt[4]{\frac{D_o^2}{L_c \times S}} \qquad C_2 = D_o\sqrt{\frac{q}{P}} \qquad C_3 = \sqrt{\frac{\delta}{L_c q}} \qquad C_4 = D_i\sqrt{\frac{q}{P}} \qquad C_5 = \sqrt[4]{\frac{D_i^2}{L_c \times S}}$$

For imperial springs values of G other than 11.5×10^6 lbf/in² C_1 and C_2 should be multiplied by

$$\sqrt{\frac{11.5 \times 10^6}{G}}$$

and C_3 multiplied by

$$\sqrt[4]{\frac{11.5 \times 10^6}{G}}$$

TABLE 8.3

Values of Design Functions C_1, C_2, C_3, C_4, and C_5 (to be used for imperial springs)

c	D_i/D_o	C_1	C_2	C_3	C_4	C_5
2.00	0.333	0.084	10.00	0.000700	3.350	0.049
3.00	0.500	0.132	14.00	0.001240	7.000	0.093
3.10	0.512	0.137	14.40	0.001290	7.380	0.097
3.20	0.524	1.142	14.90	0.001350	7.770	0.102
3.30	0.535	0.147	15.40	0.001400	8.170	0.106
3.40	0.545	0.152	15.90	0.001450	8.580	0.111

TABLE 8.3 *(Continued)*

Values of Design Functions C_1, C_2, C_3, C_4, and C_5 (to be used for imperial springs)

c	D_i/D_o	C_1	C_2	C_3	C_4	C_5
3.50	0.555	0.157	16.30	0.001500	9.012	0.116
3.60	0.565	0.162	16.80	0.001560	9.451	0.121
3.70	0.574	0.167	17.30	0.001610	9.900	0.126
3.80	0.583	0.172	17.80	0.001660	10.359	0.132
3.90	0.592	0.177	18.40	0.001710	10.826	0.137
4.00	0.600	0.183	18.90	0.001770	11.300	0.142
4.10	0.608	0.188	19.40	0.001820	11.800	0.147
4.20	0.616	0.193	19.90	0.001870	12.300	0.152
4.30	0.623	0.199	20.50	0.001920	12.800	0.157
4.40	0.630	0.204	21.00	0.001980	13.300	0.162
4.50	0.637	0.209	21.60	0.002030	13.800	0.167
4.60	0.643	0.215	22.10	0.002080	14.300	0.172
4.70	0.649	0.220	22.70	0.002130	14.800	0.177
4.80	0.655	0.225	23.30	0.002190	15.300	0.182
4.90	0.661	0.231	23.80	0.002240	15.800	0.188
5.00	0.667	0.237	24.40	0.002290	16.300	0.193
5.10	0.672	0.242	25.00	0.002340	16.800	0.198
5.20	0.677	0.248	25.60	0.002400	17.300	0.204
5.30	0.682	0.253	26.20	0.002450	17.900	0.209
5.40	0.687	0.259	26.80	0.002500	18.400	0.215
5.50	0.692	0.264	27.40	0.002550	19.000	0.220
5.60	0.697	0.270	28.00	0.002610	19.500	0.226
5.70	0.702	0.276	28.60	0.002660	20.100	0.231
5.80	0.706	0.281	29.20	0.002710	20.600	0.236
5.90	0.710	0.287	29.80	0.002760	21.200	0.242
6.00	0.714	0.293	30.40	0.002820	21.700	0.247
6.10	0.718	0.299	31.10	0.002870	22.300	0.253
6.20	0.722	0.304	31.70	0.002920	22.900	0.258
6.30	0.726	0.310	32.40	0.002970	23.500	0.264
6.40	0.730	0.316	33.00	0.003030	24.100	0.207
6.50	0.733	0.322	33.70	0.003080	24.700	0.276
6.60	0.737	0.328	34.30	0.003130	25.300	0.282
6.70	0.740	0.344	35.00	0.003180	25.900	0.287
6.80	0.744	0.340	35.70	0.003240	26.500	0.293
6.90	0.747	0.346	36.30	0.003290	27.100	0.299
7.0	0.750	0.352	37.00	0.00334	27.70	0.305
7.1	0.753	0.358	37.70	0.00339	28.40	0.311
7.2	0.756	0.364	38.40	0.00344	29.00	0.316
7.3	0.759	0.370	39.00	0.00350	29.60	0.322
7.4	0.762	0.376	39.70	0.00355	30.20	0.328

TABLE 8.3 *(Continued)*
Values of Design Functions C_1, C_2, C_3, C_4, and C_5
(to be used for imperial springs)

c	D_i/D_o	C_1	C_2	C_3	C_4	C_5
7.5	0.765	0.382	40.40	0.00360	30.90	0.334
7.6	0.767	0.388	41.10	0.00365	31.50	0.340
7.7	0.770	0.394	41.80	0.00371	32.20	0.346
7.8	0.773	0.400	42.50	0.00376	32.80	0.351
7.9	0.775	0.406	43.20	0.00381	33.50	0.357
8.0	0.778	0.412	44.00	0.00386	34.20	0.363
8.1	0.780	0.418	44.70	0.00392	34.90	0.369
8.2	0.783	0.424	45.40	0.00397	35.50	0.375
8.3	0.785	0.431	46.10	0.00402	36.20	0.382
8.4	0.787	0.437	46.90	0.00407	36.90	0.388
8.5	0.789	0.443	47.60	0.00412	37.60	0.394
8.6	0.792	0.449	48.30	0.00418	38.30	0.400
8.7	0.794	0.456	49.10	0.00423	39.00	0.406
8.8	0.796	0.462	49.80	0.00428	39.70	0.412
8.9	0.798	0.468	50.60	0.00433	40.40	0.418
9.0	0.800	0.474	51.30	0.00439	41.10	0.424
9.1	0.802	0.481	52.10	0.00444	41.82	0.430
9.2	0.804	0.487	52.90	0.00449	42.55	0.437
9.3	0.806	0.494	53.60	0.00454	43.29	0.443
9.4	0.808	0.500	54.40	0.00460	44.03	0.449
9.5	0.810	0.506	55.20	0.00465	44.77	0.456
9.6	0.811	0.513	55.90	0.00470	45.51	0.462
9.7	0.813	0.519	56.70	0.00475	46.25	0.468
9.8	0.815	0.526	57.50	0.00480	46.99	0.475
9.9	0.817	0.532	58.30	0.00486	47.70	0.481
10.0	0.818	0.539	59.10	0.00491	48.40	0.487
10.1	0.820	0.545	59.90	0.00496	49.15	0.493
10.2	0.821	0.552	60.70	0.00501	49.90	0.499
10.3	0.823	0.558	61.50	0.00507	50.65	0.505
10.4	0.825	0.565	62.30	0.00512	51.41	0.512
10.5	0.826	0.571	63.10	0.00517	52.16	0.518
10.6	0.828	0.578	63.90	0.00522	52.92	0.524
10.7	0.829	0.584	64.70	0.00528	53.69	0.531
10.8	0.831	0.591	65.50	0.00533	54.45	0.537
10.9	0.832	0.598	66.40	0.00538	55.23	0.544
11.0	0.833	0.604	67.20	0.00543	56.00	0.551
12.0	0.846	0.671	75.70	0.00596	64.10	0.617

Values of design functions, C_1, C_2, C_3, C_4, and C_5.

$$C_1 = \sqrt[4]{\frac{D_o^2}{L_c \times S}} \qquad C_2 = D_o\sqrt{\frac{q}{P}} \qquad C_3 = \sqrt{\frac{\delta}{L_c \times q}} \qquad C_4 = D_i\sqrt{\frac{q}{P}} \qquad C_5 = \sqrt[4]{\frac{D_i^2}{L_c \times S}}$$

For metric springs values of G other than 79.3 MPa C_1 and C_2 should be multiplied by

$$\sqrt{\frac{79.3 \text{ MPa}}{G}}$$

and C_3 multiplied by

$$\sqrt[4]{\frac{79.3 \text{ MPa}}{G}}$$

TABLE 8.4

Values of Design Functions C_1, C_2, C_3, C_4, and C_5 (to be used for metric springs (SI Units))

c	D_i/D_o	C_1	C_2	C_3	C_4	C_5
2.00	0.333	0.292	10.00	0.0084	3.350	0.170
3.00	0.500	0.458	14.00	0.0149	7.000	0.323
3.10	0.512	0.475	14.40	0.0155	7.380	0.337
3.20	0.524	0.493	14.90	0.0163	7.770	0.354
3.30	0.535	0.510	15.40	0.0169	8.170	0.368
3.40	0.545	0.527	15.90	0.0175	8.580	0.385
3.50	0.555	0.545	16.30	0.0181	9.012	0.403
3.60	0.565	0.562	16.80	0.0188	9.451	0.420
3.70	0.574	0.580	17.30	0.0194	9.900	0.437
3.80	0.583	0.597	17.80	0.0200	10.359	0.458
3.90	0.592	0.614	18.40	0.0206	10.826	0.475
4.00	0.600	0.635	18.90	0.0213	11.300	0.493
4.10	0.608	0.652	19.40	0.0219	11.800	0.510
4.20	0.616	0.670	19.90	0.0225	12.300	0.527
4.30	0.623	0.691	20.50	0.0231	12.800	0.545
4.40	0.630	0.708	21.00	0.0238	13.300	0.562
4.50	0.637	0.725	21.60	0.0244	13.800	0.580
4.60	0.643	0.746	22.10	0.0250	14.300	0.597
4.70	0.649	0.763	22.70	0.0257	14.800	0.614
4.80	0.655	0.781	23.30	0.0264	15.300	0.632
4.90	0.661	0.802	23.80	0.0270	15.800	0.652
5.00	0.667	0.822	24.40	0.0276	16.300	0.670
5.10	0.672	0.840	25.00	0.0282	16.800	0.687
5.20	0.677	0.861	25.60	0.0289	17.300	0.708
5.30	0.682	0.878	26.20	0.0295	17.900	0.725
5.40	0.687	0.899	26.80	0.0301	18.400	0.746

TABLE 8.4 *(Continued)*
Values of Design Functions C_1, C_2, C_3, C_4, and C_5 (to be used for metric springs (SI Units))

c	D_i/D_o	C_1	C_2	C_3	C_4	C_5
5.50	0.692	0.916	27.40	0.0307	19.000	0.763
5.60	0.697	0.937	28.00	0.0314	19.500	0.784
5.70	0.702	0.958	28.60	0.0320	20.100	0.802
5.80	0.706	0.975	29.20	0.0326	20.600	0.819
5.90	0.710	0.996	29.80	0.332	21.200	0.840
6.00	0.714	1.017	30.40	0.0340	21.700	0.857
6.10	0.718	1.038	31.10	0.0346	22.300	0.878
6.20	0.722	1.055	31.70	0.0352	22.900	0.892
6.30	0.726	1.076	32.40	0.0358	23.500	0.916
6.40	0.730	1.097	33.00	0.0365	24.100	0.718
6.50	0.733	1.117	33.70	0.0371	24.700	0.958
6.60	0.737	1.138	34.30	0.0377	25.300	0.979
6.70	0.740	1.194	35.00	0.383	25.900	0.996
6.80	0.744	1.180	35.70	0.0390	26.500	1.017
6.90	0.747	1.201	36.30	0.0396	27.100	1.038
7.0	0.750	1.222	37.00	0.0402	27.70	1.058
7.1	0.753	1.242	37.70	0.0408	28.40	1.079
7.2	0.756	1.263	38.40	0.0414	29.00	1.097
7.3	0.759	1.284	39.00	0.0422	29.60	1.117
7.4	0.762	1.305	39.70	0.0428	30.20	1.138
7.5	0.765	1.326	40.40	0.0434	30.90	1.159
7.6	0.767	1.346	41.10	0.0440	31.50	1.180
7.7	0.770	1.367	41.80	0.0447	32.20	1.201
7.8	0.773	1.388	42.50	0.0453	32.80	1.218
7.9	0.775	1.409	43.20	0.0459	33.50	1.239
8.0	0.778	1.430	44.00	0.0465	34.20	1.260
8.1	0.780	1.451	44.70	0.0472	34.90	1.281
8.2	0.783	1.471	45.40	0.0478	35.50	1.301
8.3	0.785	1.496	46.10	0.0484	36.20	1.326
8.4	0.787	1.517	46.90	0.0490	36.90	1.346
8.5	0.789	1.537	47.60	0.0496	37.60	1.367
8.6	0.792	1.558	48.30	0.0503	38.30	1.388
8.7	0.794	1.582	49.10	0.0509	39.00	1.409
8.8	0.796	1.603	49.80	0.0515	39.70	1.430
8.9	0.798	1.624	50.60	0.0521	40.40	1.451
9.0	0.800	1.645	51.30	0.0529	41.10	1.471
9.1	0.802	1.669	52.10	0.0535	41.82	1.492
9.2	0.804	1.690	52.90	0.0541	42.55	1.517
9.3	0.806	1.714	53.60	0.0547	43.29	1.537
9.4	0.808	1.735	54.40	0.0554	44.03	1.558

TABLE 8.4 *(Continued)*
Values of Design Functions C_1, C_2, C_3, C_4, and C_5
(to be used for metric springs (SI Units))

c	D_i/D_o	C_1	C_2	C_3	C_4	C_5
9.5	0.810	1.756	55.20	0.0560	44.77	1.582
9.6	0.811	1.780	55.90	0.0566	45.51	1.603
9.7	0.813	1.801	56.70	0.0572	46.25	1.624
9.8	0.815	1.825	57.50	0.0578	46.99	1.648
9.9	0.817	1.846	58.30	0.0585	47.70	1.669
10.0	0.818	1.871	59.10	0.0591	48.40	1.690
10.1	0.820	1.891	59.90	0.0597	49.15	1.711
10.2	0.821	1.916	60.70	0.0603	49.90	1.732
10.3	0.823	1.936	61.50	0.0611	50.65	1.753
10.4	0.825	1.961	62.30	0.0617	51.41	1.777
10.5	0.826	1.982	63.10	0.0623	52.16	1.798
10.6	0.828	2.006	63.90	0.0629	52.92	1.818
10.7	0.829	2.027	64.70	0.0636	53.69	1.843
10.8	0.831	2.051	65.50	0.0642	54.45	1.864
10.9	0.832	2.075	66.40	0.0648	55.23	1.888
11.0	0.833	2.096	67.20	0.0654	56.00	1.912
12.0	0.846	2.329	75.70	0.0718	64.10	2.141

TABLE 8.5
Number of Ineffective Coils
for Various End Coil Forms

End Coil Formation	Total Number of Ineffective Coils (N–n)
One coil at each end closed and ground	1.5–2.0
End coils unclosed but ground	0.5–1.0
End coils unclosed and unground	0–0.5
One coil at each end closed but unground	1.5–2.0

8.1.6   SPRING AND DEFLECTION CHARACTERISTICS

The axial spring rate, S, of a compression spring may be calculated from Equation (8.2).

Data on the transverse spring rate, S_t, is presented in Section 8.1.9.

$$S = \frac{G \times d}{8 \times n \times c^3} \tag{8.2}$$

The relationship between N and n is shown in Table 8.5. The value of N – n represents the total number of ineffective coils at both ends of the spring. In each range, the higher number corresponds to the compression under a large load and the lower number to compression under a small load.

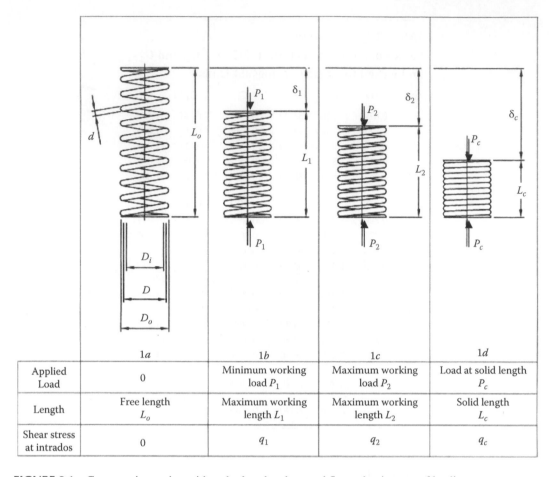

	1a	1b	1c	1d
Applied Load	0	Minimum working load P_1	Maximum working load P_2	Load at solid length P_c
Length	Free length L_o	Maximum working length L_1	Maximum working length L_2	Solid length L_c
Shear stress at intrados	0	q_1	q_2	q_c

FIGURE 8.1 Compression spring with ends closed and ground flat under 4 stages of loading.

8.1.7 SOLID LENGTH CHARACTERISTICS

The nominal solid length of a spring is the length in which all coils are touching (see Figure 8.1). Some additional notes on the solid length are covered in Section 8.3.

Most springs are scragged to introduce favorable residual stresses into the wire. If advantage of this process is not taken, then it will be necessary to ensure the spring will not be overstressed when compressed solid. At the minimum working length, clearance should be maintained between the coils. More detailed notes are contained in Section 8.3.

The nominal length of the spring will vary dependent upon the type of end coil formation and total number of coils in the spring, N. The relationship between n and N is presented in Table 8.5, and expressions for the nominal solid length are given in Table 8.6.

8.1.8 BUCKLING OF COMPRESSION SPRINGS

A compression spring will buckle if the deflection under a maximum force, as a proportion of the free length of the spring, exceeds a critical value of HD/L_o. A range of values of H for various end conditions is shown in Figure 8.2, and critical values of HD/L_o are shown in Figure 8.3.

There are two main reasons why a range of values for H has been given for each end condition:

1. The exact end condition as shown in Figure 8.2 is rarely met in actual practice.
2. Buckling is affected to a large extent by the squareness of the end of the spring under load.

TABLE 8.6
Effect of End Coil Forms on the Solid Length of the Spring

End Coil Formation	L_c
One coil at each end closed and ground	d(N – 0.5)[a]
One coil at each end closed but unground	d(N + 1)
End coils unclosed and unground	d(N + 1)
End coils unclosed and ground	d(N – 0.5)[a]

[a] This expression assumes that the tip of the end coil has been ground down to a thickness of 0.25d. That is, in the case of one coil at each end closed and ground, ¾ of the end coil is ground.

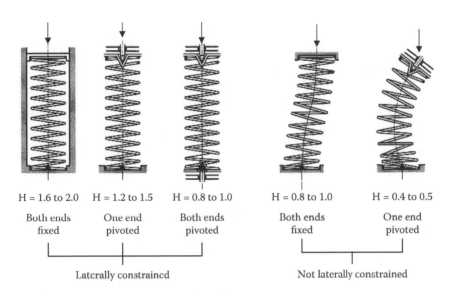

FIGURE 8.2   End function "H" for various end conditions.

For the purposes of calculation the minimum value of H is used.

8.1.9   TRANSVERSE LOADING

In some designs the compressed spring may be subjected to a combined horizontal force as depicted in Figure 8.4.

The transverse rate of the spring with the end coils restrained to remain parallel to each other is given by Equation (8.3).

$$(S_t)_{p=0} = \frac{2S(1+v)}{1+\frac{(2+v)}{3}\left(\frac{L_o}{D}\right)^2} \tag{8.3}$$

In most instances, the spring may be subjected to an axial load and the transverse rate is then given by Equation (8.4).

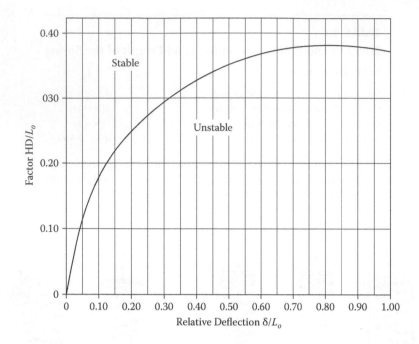

FIGURE 8.3   Critical values of HD/L₀.

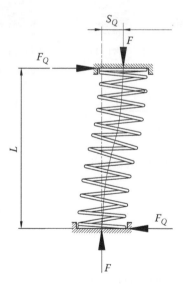

FIGURE 8.4   Compression spring under combined axial and transverse loading.

$$S_t = \frac{S\left(\dfrac{L_o}{L} - 1\right)}{\left[\left(1 + \dfrac{\left(\dfrac{L_o}{L} - 1\right)}{2(1+\nu)}\right)\dfrac{\tan k}{k} - 1\right]} \tag{8.4}$$

where
$$k = \frac{L}{2D}\left[2\left(\frac{L_o}{L}-1\right)\left(\frac{2+v}{1+v}\right)\left(1+\frac{\left(\frac{L_o}{L}-1\right)}{2(1+v)}\right)\right]^{0.5}$$

8.1.10   INCREASE OF SPRING DIAMETER UNDER COMPRESSION

The coil diameter of a compression spring will increase when the spring is compressed. In general this expansion is within the region of 2%. The expansion will be dependent on whether the end coils are free to rotate about the helix angle.

   The following derivation has been developed from Wahl.

8.1.10.1   Ends Free to Rotate

The relative change in the diameter of the spring, ΔD, during compression from the free length, L_o, to the solid length, L_c, is given by

$$\frac{\Delta D}{D} = \frac{1.3\,p^2 - 0.3\,d^2 - pd}{13\,D^2} \tag{8.5}$$

where the unloaded pitch,

$$p = \frac{L_o - \bar{d}}{n_0} \tag{8.6}$$

8.1.10.2   Ends Restrained against Rotation

The relative change in the diameter during compression from free length, L_o, to solid length, L_c, is given by

$$\frac{\Delta D}{D} = \left(1 + \frac{p^2 - d^2}{\pi^2 \, x \, D^2}\right)^{1/2} - 1 \tag{8.7}$$

   In reality, the actual coil expansion will be between the values derived from Equations (8.5) and (8.7).

8.1.11   HELIX WARPING IN COMPRESSION SPRINGS

Various assumptions have been made in Section 8.1 about the deflection behavior of helical compression springs. Wahl (1953) gives a more detailed examination and shows that these assumptions lead to an underestimation of the stress, particularly in short springs.

   Figure 8.5 gives the relationship between q_w/q and the stress increase due to helix warping and N for springs with one coil at each end closed and ground flat.

8.1.12   NATURAL FREQUENCY

The majority of helical compression springs are subject to dynamic applications such as:

- Restraining a cam follower such as an inlet and exhaust valve
- Absorbing shock or deaccelerating a load
- Providing an accelerating force to a load

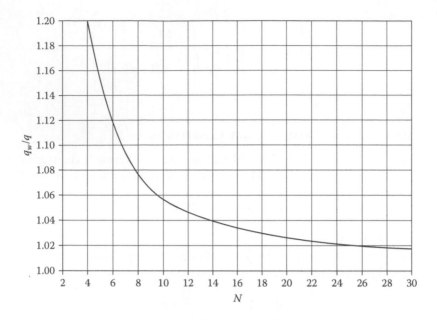

FIGURE 8.5 Stress increase due to helix warping.

All springs have their own natural frequency, and it is the responsibility of the engineer to ensure that the spring will operate well under its first natural frequency; there will be a risk of damage, say in the case of an inlet or exhaust valve in an engine colliding with the piston if the spring cannot return the valve in time.

8.1.12.1 First Natural Frequency

A natural frequency curve is shown in Figure 8.6. The first natural frequency for a helical spring that is fixed at both ends is found to be

$$f_{nat} = \frac{d}{9D^2 n_t} \sqrt{\frac{G}{\rho}} \tag{8.8}$$

where
  d = wire diameter
  D = nominal coil diameter
  n_t = total number of coils
  G = shear modulus
  ρ = material density

Figure 8.7 shows typical stresses induced in a spring that is subjected to various cyclic loading.
Springs subjected to resonant forcing frequencies do exhibit a degree of damping. Figure 8.8 shows a typical damping curve for steel springs.

8.1.13 EXAMPLE 1

8.1.13.1 Design of a Helical Compression Spring

A compression spring is to provide a load of 30 N when compressed to a length of 50 mm and a load of 85 N when compressed a further 20 mm. The outside diameter must not exceed 32 mm.
Both ends of the spring are fixed and laterally guided.

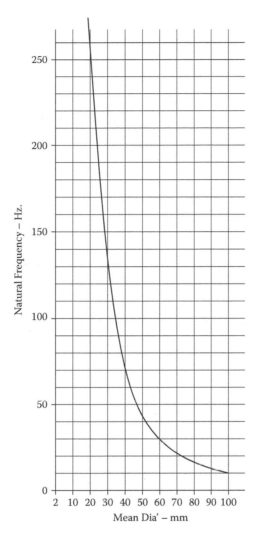

FIGURE 8.6 Natural frequency curve.

The spring shall have a fatigue life in excess of 10^7 cycles when driven by a cam rotating at 500 rev/min (8.33 Hz).

From the above information Figure 8.9 is produced.

8.1.13.1 End Coil Formulation

In this case it was considered that one coil at each end closed and ground flat would be appropriate.

8.1.13.2 Working Stress

A value of 500 MN/m² was chosen for the static design shear stress.

8.1.13.3 Determination of D, d, and c

The coil growth under compression is not considered to exceed 2% of the diameter, and allowing for the possible maximum tolerance on the outside diameter, a value of $D_o = 30$ mm is considered appropriate.

$$\frac{P}{q} = \frac{85N}{500 \times 10^6 N/m^2}$$

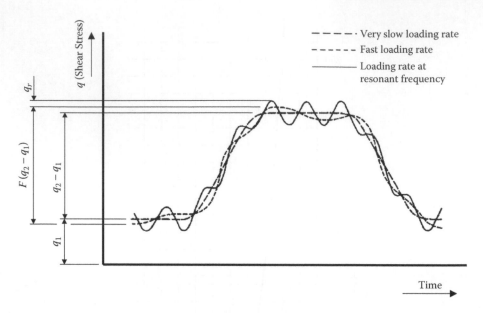

FIGURE 8.7 Stresses in spring subject to various cyclic loading.

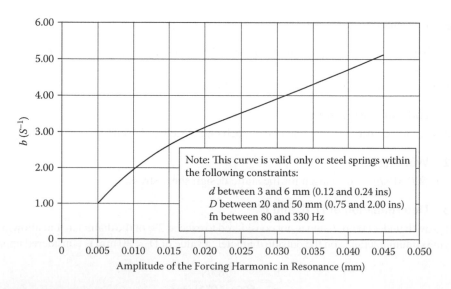

FIGURE 8.8 Damping coefficient.

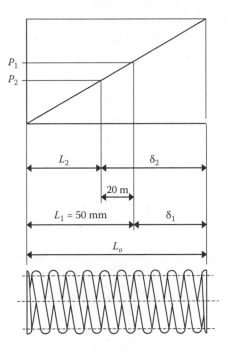

FIGURE 8.9   Compression spring diagram.

At maximum working load:
$$= 0.17 \times 10^{-6} \text{ m}^2$$

Considering case 2 from Table 8.3,

$$C_2 = D_o \sqrt{\frac{q}{P}}$$

$$= 30 \times 10^{-3} \sqrt{\frac{500 \times 10^6}{85}}$$

$$= 72.76$$

From Table 8.4 the spring index c corresponding to a value of $C_2 = 72.76$:
$$= 11.654 \text{ (by interpolation)}$$

Corresponding wire diameter d is calculated from

$$d = \frac{D_o}{c+1}$$

$$= \frac{30 \times 10^{-3}}{11.654 + 1}$$

$$= 2.371 \times 10^{-3} \text{ m}$$

$$\text{say } 2.371 \text{ mm}$$

As this wire diameter does not correspond to a standard wire size, a wire diameter of 2.50 mm will be chosen for this calculation. Choosing a larger wire size will reduce the working stress.

Hence recalculating the spring index c:

$$c = \frac{(D_o - d)}{d}$$

$$= \frac{30 - 2.5}{2.5}$$

$$c = 11$$

and

$$D = 30 - 2.5$$
$$= 27.5 \text{ mm}$$

Recalculating the working stress in the spring at the maximum working load:

$$q = \frac{8PD}{\pi d^3} \times \frac{c + 0.2}{c - 1}$$

$$= \frac{8 \times 85 \times 27.5}{\pi \times 2.5^3}$$

$$= 426.67 \text{ MPa}$$

8.1.13.4 Determining Spring Rate, Deflection Characteristics, and Number of Coils

$$S = \frac{\text{Final load} - \text{initial load}}{\text{Deflection from initial load to final load}}$$

$$= \frac{85 \text{ N} - 30 \text{ N}}{(50 - 30) \times 10^{-3} \text{m}}$$

$$S = 2750 \text{ N/m}$$

Assuming a value of G for steel of 79.3 GN/m²,
Considering the number of working coils:

$$S = \frac{Gd}{8nc^3}$$

$$n = \frac{Gd}{8Sc^3}$$

$$= \frac{79.3 \times 10^9 \times 2.5 \times 10^{-3}}{8 \times 2750 \times 11^3}$$

$$= 6.77 \text{ coils}$$

From Table 8.5 for springs with ends closed and ground flat:

$$N - n = 1.75 \text{ (mid-range)}$$

Therefore,

$$N = 1.75 + 6.77$$
$$= 8.52 \text{ total coils}$$

The initial deflection (δ_1) produced by the initial load of 30 N is

$$\delta_1 = \frac{P}{S}$$

$$= \frac{30\,N}{2750 \text{ N/m}}$$

$$\delta_1 = 10.91 \text{ mm}$$

and the free length (L_o)

$$L_0 = \delta_1 + L_1$$

$$= 10.91 \text{ mm} + 50 \text{ mm}$$

$$= 60.91 \text{ mm}$$

8.1.13.5   Conformity with Limitations of Section 8.3

From Equation 8.1 and Table 8.1, together with the assumptions of Section 8.1.13.1 that one coil at each end is closed and ground,

$$n_o = N - 2$$

$$L_o < 0.55 n_o\, D + \overline{d}$$

Since $L_o = 60.91$ mm is less than 102.8 mm, the helix angle of the unloaded spring is less than 10°.

Furthermore, the deflection is given by

$$0.55 n_0\, D + \overline{d} = 0.55 \times (N - 2) \times D + 1.5\,d$$

$$= 0.55 \times 6.55 \times 27.5 + 1.5 \times 2.5$$

$$= 102.82 \text{ mm}$$

This is less than

$$\frac{D}{4}\left(\frac{27.5}{4}\right) = 6.88 \text{ mm}$$

and so both criteria of Section 8.1.3 are satisfied.

8.1.13.6   Closed Length Characteristics

$$\frac{\delta_2}{n_0} = \frac{10.91 \text{ mm} + 20.0 \text{ mm}}{6.55}$$

$$= 4.72 \text{ mm}$$

Using Table 8.6, the nominal solid length is given by

$$L_c = d(N - 0.5)$$

$$= 2.5 \text{ mm } (8.52 - 0.5)$$

$$= 20.05 \text{ mm}$$

Since the minimum working length is 30 mm, the spring does not approach the solid length when working. If the working length was only slightly larger than the nominal solid length, it would be necessary to compare the dimensional tolerances on the working length with the maximum solid length.

The additional load necessary to close the spring to its solid length is given by

$$(L_2 - L_c) \times S = (30 \times 10^{-3} - 20.05 \times 10^{-3} \times 2750$$

$$= 27.36 \text{ N}$$

Therefore, the total load needed to close the spring is

$$= 27.36 \text{ N} + 85 \text{ N}$$

$$= 112.36 \text{ N}$$

and the shear stress at closure:

$$= \frac{112.36}{\left(\dfrac{P}{q} \right)}$$

$$= \frac{112.36 \times 462.39}{85}$$

$$= 611.225 \text{ MN} / \text{m}^2$$

This stress would be permissible for a spring manufactured from BS 2803, which is low temperature, heat treated, and prestressed.

8.1.13.7 Buckling Characteristics

For a nonpivoted spring, as both ends of the spring are fixed and laterally constrained, from Figure 8.2 H has the values of 1.6 to 2.0.

In practice, however, H should be reduced by 20%, so H is taken to have a value of 1.6. Therefore,

$$\frac{HD}{L_0} = \frac{1.6 \times 27.5}{60.91}$$

$$= 0.722$$

$$\frac{Ha}{L_0} = \frac{1.6 \times 0}{60.91}$$

$$= 0$$

From Figure 8.3, where HD/L_0 is greater than 0.4, the spring will be stable for any deflection.

8.1.13.8   Determination of Coil Growth under Compression

From Equation 8.6,

$$p = \frac{L_o - \bar{d}}{n_o}$$

$$= \frac{60.91 - 1.5 \times 2.5}{6.55}$$

$$= 8.727 \text{ mm}$$

For a worst case, that is, ends free to rotate, Equation (8.5) applies:

$$\frac{\Delta D}{D} = \frac{1.3p^2 - 0.3d^2 - pd}{13D^2}$$

$$= \frac{1.3 \times 8.727^2 - 0.3 \times 2.5^2 - 8.727 \times 2.5}{13 \times 27.5^2}$$

$$= 0.00766$$

Thus, the increase in the diameter of the spring is given by

$$\Delta D = 0.00766 \times 27.5 \text{ mm}$$

$$= 0.211 \text{ mm}$$

The maximum outside diameter of the spring will be 30.21 mm.

It is recommended to increase the outside diameter to 31.0 mm to cover any variations of manufacturing tolerances.

8.1.13.9   Natural Frequency of the Spring

From Equation (8.8) the natural frequency of the spring is calculated as follows:

$$\omega_n = \frac{d}{9D^2 n_t} \sqrt{\frac{G}{\rho}}$$

where:

d = 2.5 mm
D = 27.5 mm
n_t = 8.52
G = 79.3 $\times 10^9$ Pa
ρ = 7,850 kg/m^3

hence,

ω_n = 137 Hz

In this example the operating frequency of the spring is 500 rev/min (8.33 Hz), which is well below the natural frequency of the spring. Therefore, in this example there will be no significant increase in the stress in the spring.

8.1.13.10 Stress Increase Due to Helix Warping

From Figure 8.5 with N = 8.52, q_w/q = 1.068, this indicates that the stress in the spring previously calculated is some 6% low. Thus, the stress at static working load is

$$q = 425 \text{ MN/m}^2 \times 1.068$$

$$= 453.9 \text{ MN/m}^2$$

Similarly, the surge stress will be

$$= 561.77 \text{ MN/m}^2$$

The closure stress will not be affected by helix warping.

8.2 ALLOWABLE STRESSES FOR HELICAL COMPRESSION SPRINGS MANUFACTURED FROM ROUND WIRE

Section 8.2 provides data on both the static and fatigue strengths of various spring materials. The data relate to prestressed springs that have been heat treated after manufacture in cases where this has a beneficial effect on the fatigue life of the spring and operating in a noncorrosive ambient environment. For most carbon steel spring materials, an ambient temperature is within the range of −20 to +150°C.

In general the static and fatigue strengths should be checked in the design. Static criteria alone may be used for springs that are limited to a few hundred cycles in their working life.

8.2.1 STATIC STRENGTH DATA

Once the major parameters are established, the stress when the spring is compressed to solid may be calculated.

Usually, one of the first steps in spring design is to choose a value of stress that corresponds to the maximum working load. For a spring that only sustains a static load, this should be as high as reasonably possible, normally about 85% of the limiting static shear stress of the spring material.

This should give an adequate clearance between the coils when the spring is compressed to its minimum working length.

The shear stress q in Figures 8.10 and 8.11 refers to springs that have a degree of prestress. Prestressing allows springs to be loaded in service beyond the point that ordinarily produces plastic deformation and therefore a permanent set. Figures 8.10 and 8.11 show the stress distribution during the prestressing process. The compression spring is initially wound to be longer than required and is then stressed elastically under a compressive force from P to Q (Figure 8.10) and then plastically to closure from Q to R.

Releasing the force leads to a reverse residual stress to be generated, $-q_s$. Subsequently, the spring is able to accept a change of stress, $|q_s| + |q_r|$, while remaining elastic. In practice the spring is compressed to solid five or six times (S to T) to complete the process.

8.2.2 FATIGUE DATA

Fatigue data are generally presented on modified Goodman diagrams as in Figure 8.12(a). This figure shows how these stresses are presented on the modified Goodman-type diagram and their relationship to the working stress cycle of the spring is shown in Figure 8.12(b).

These diagrams are specifically generated for springs of a particular wire diameter and material relating to springs that have been prestressed.

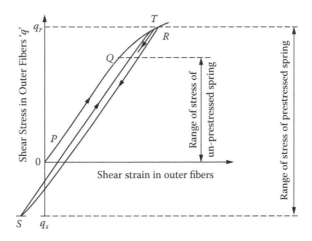

FIGURE 8.10   Shear stress in outer fibers of spring wire during prestressing.

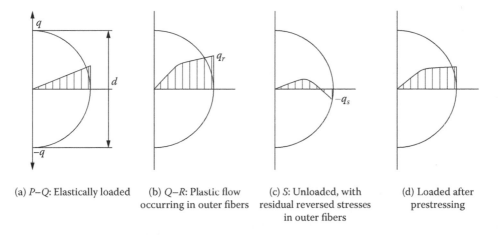

(a) P–Q: Elastically loaded   (b) Q–R: Plastic flow occurring in outer fibers   (c) S: Unloaded, with residual reversed stresses in outer fibers   (d) Loaded after prestressing

FIGURE 8.11   Shear stress distribution before, during, and after prestressing superimposed on wire diameter.

8.2.2.1   Estimation of an S-N Curve

Figure 8.13 shows a typical S-N curve; the stress amplitude (Sa) is plotted against the number of cycles to failure (N). The curve indicates the number of cycles to failure of the material, for a given specific stress amplitude, between 0 and 10^7 cycles. N is usually plotted on a logarithmic scale because the value may range between a few thousand to a few million cycles. The stress may also be plotted on a logarithmic scale. The line represents the best fit for the data points generated from the numerous fatigue tests undertaken. The fatigue test may have been a straight push-pull type of test or a bend test based on the Wohler rotating beam test.

The curve in Figure 8.13 is for a material manufactured from BS 1408 range 3 (unpeened carbon steel) subject to a constant mean stress of 500 MN/m².

8.2.3   Factors Affecting Spring Life

8.2.3.1   Spring Geometry

At small values of the spring index, c, the shear stress at the inner surface of the coil is significantly increased due to the curvature of the wire. This effect is accounted for in the curvature

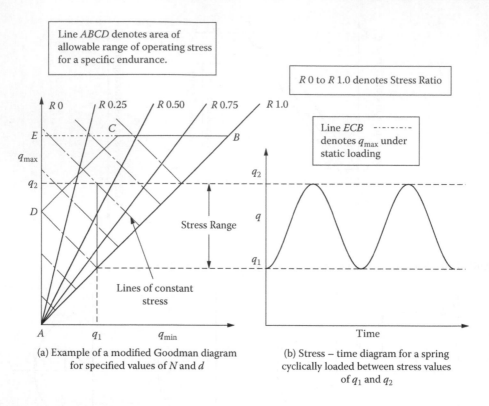

Line *ABCD* denotes area of allowable range of operating stress for a specific endurance.

$R\,0$ to $R\,1.0$ denotes Stress Ratio

Line *ECB* ---------- denotes q_{max} under static loading

Stress Range

Lines of constant stress

(a) Example of a modified Goodman diagram for specified values of N and d

(b) Stress – time diagram for a spring cyclically loaded between stress values of q_1 and q_2

FIGURE 8.12   Presentation of fatigue data.

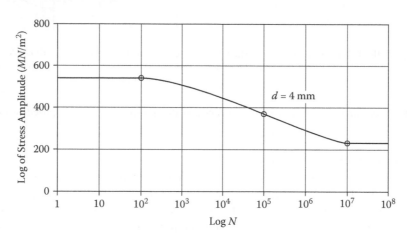

$d = 4$ mm

FIGURE 8.13   Estimated S-N curve for BS 1408 Range 3 at a constant mean stress: 500MN/m².

correction factor (see Section 8.1.5.2) when calculating q. Springs coiled from small-diameter wire can withstand higher operating stresses than larger wire diameters. There are two reasons for this:

1. A greater reduction in wire diameter by drawing during the manufacturing process improves the mechanical properties.
2. The springs may be coiled from cold wire. If hot coiled springs are subsequently heat treated, then decarburation problems may result.

8.2.3.2   Corrosion

The fatigue life of the spring may be significantly reduced if the spring is operating in a corrosive atmosphere. The fatigue life could reduce to as little as 10% of the fatigue life. Plating or other anti-corrosion treatments may protect springs, but care will be required to ensure the treatment will not have a deleterious effect on the life of the spring due to, say, hydrogen embrittlement.

Corrosion-resistant materials such as titanium, stainless steel, or copper-nickel alloys would be a better choice of material.

8.2.3.3   Surface Finish

Machine marks or other surface damage on the surface of the spring may lead to early fatigue failure, as these will allow crack initiation to start much easier. A good surface finish can be obtained by specifying valve spring quality wire.

8.2.3.4   Elevated Temperatures

High operating temperatures tend to induce a permanent set or loss of load in springs at a fixed deflection and a reduction in the shear modulus. Temperatures as low as 120°C may cause a relaxation of, for instance, music wire (BS 1408M, ASTM A228, etc.), although other materials can withstand higher temperatures.

8.2.4   Treatments for Improving the Fatigue Life of Springs

8.2.4.1   Prestressing

Prestressing is a process where the compression spring is cycled between its operating range several times to improve the static and fatigue properties. It is described in more detail in Section 8.3.2.

8.2.4.2   Shot Peening

Shot peening is a process where small pellets are fired at the surface of the spring under very high pressure. It has the most beneficial effect where the torsional stress is greatest on the inside of the coils. Shot peening is difficult to accomplish with springs of small diameter wire and close coiled springs.

After shot peening the spring should be given a low-temperature heat treatment to further enhance its fatigue life. Peening leaves a very active surface, which should be protected against corrosion.

The process improves the fatigue life in three ways:

1. The process induces compressive stresses into the outer skin of the spring surface, thereby reducing the overall effects of the tensile stresses at the surface.
2. Part of the improvement in fatigue life due to shot peening may be attributed to increasing the surface hardness. It should be noted that shot peening has less effect on harder wires.
3. The process improves the surface finish and can remove small surface imperfections; the small indentations left by the shot will have no deleterious effects on crack propagation.

The process should not be relied on to improve very poor or decarburized surfaces.

8.2.4.3   Abrasive Cleaning

Abrasive cleaning improves the fatigue life by smoothing the surface of the spring, although it has the disadvantage of removing material from the wire. It is accomplished by blasting abrasive particles at the surface using compressed air or water. The surface should be immediately protected against corrosion.

TABLE 8.7

Spring Wire Data: Imperial Sizes and Allowables

Dimensions used:
Temperature: Degrees Centigrade
Strength: Lbf/in.^{2}
Modulus: Lbf/in.^{2}
Density: Lbf/in.^{3}

Material Wire Diameter	Stainless Steel S205	Steel S201 or 202	Steel Def. 106	Nimonic 90 HR501 HD	Nimonic 90 HR502	HD Phos. Brz. BS 384
0.3000	63,616	62,720	85,120	76,000	62,650	49,250
0.2760	63,616	62,720	85,120	76,000	62,650	49,250
0.2520	69,888	71,680	89,600	76,000	62,650	49,250
0.2320	69,888	71,680	89,600	76,000	62,650	49,250
0.2120	72,576	71,680	89,600	76,000	62,650	49,250
0.1920	75,266	80,640	94,080	80,640	62,650	49,250
0.1760	77,952	80,640	94,080	80,640	62,650	49,250
0.1600	80,640	80,640	94,080	80,640	62,650	49,250
0.1440	84,224	80,640	98,560	80,640	62,650	49,250
0.1280	86,910	89,600	98,560	80,640	62,650	49,250
0.1160	89,600	89,600	103,040	80,640	62,650	49,250
0.1040	89,600	89,600	107,520	80,640	62,650	52,000
0.0920	92,288	89,600	116,480	80,640	62,650	52,000
0.0800	95,872	98,560	116,480	80,640	62,650	52,000
0.0720	98,560	98,560	116,480	80,640	62,650	52,000
0.0640	98,560	98,560	125,440	80,640	62,650	52,000
0.0560	101,248	98,560	125,440	80,640	62,650	52,000
0.0480	103,936	107,520	125,440	80,640	62,650	52,000
0.0400	103,936	107,520	129,920	80,640	62,650	52,000
0.0360	107,520	107,520	134,400	89,340	62,650	52,000
0.0320	110,208	116,480	138,880	89,340	62,650	52,000
0.0280	112,896	116,480	143,360	89,340	62,650	52,000
0.0240	116,480	116,480	147,840	89,340	62,650	52,000
0.2200	116,480	125,440	147,840	89,340	62,650	52,000
0.0200	119,168	125,440	152,320	89,340	62,650	52,000
0.0180	119,168	125,440	152,320	89,340	62,650	0
0.0164	119,168	125,440	152,320	0	0	0
0.0148	121,856	125,440	152,320	0	0	0
0.0136	121,856	125,440	152,320	0	0	0
0.0124	121,856	125,440	152,320	0	0	0
0.0116	124,544	125,440	152,320	0	0	0
0.0108	124,544	134,400	152,320	0	0	0
0.0100	124,544	134,400	152,320	0	0	0
0.0092	124,544	134,400	0	0	0	0
0.0084	124,544	134,400	0	0	0	0
0.0076	124,544	134,400	0	0	0	0
Torsional modulus	10.00×10^6	12.00×10^6	12.00×10^6	12.00×10^6	12.00×10^6	12.00×10^6	6.50×10^6
Bending modulus	29.00×10^6	29.00×10^6	29.00×10^6	29.00×10^6	29.00×10^6	29.00×10^6	18.00×10^6
Density	0.283	0.283	0.283	0.283	0.300	0.300	0.320

TABLE 8.8
Spring Wire Data: Metric Sizes and Allowable Stresses

Dimensions used:
Temperature
Degrees Centigrade
Strength: $N/m\,m^2$
Modulus: $N/m\,m^2$
Density: $N/m\,m^3$

Material Wire Diameter	Stainless Street S205	Steel S201 or 202	Steel Def. 106	Nimonic 90 HR501 HD	Nimonic 90 HR502	HD Phos. Brz. BS 384
12.000	400	431	567	400	432	340
9.000	400	431	567	400	432	340
8.000	440	431	567	400	432	340
7.100	440	431	567	400	432	340
6.300	480	494	618	400	432	340
5.600	500	494	618	400	432	340
5.000	520	557	618	556	432	340
4.500	540	557	649	556	432	340
4.000	560	557	649	556	432	340
3.550	580	557	680	556	432	340
3.150	600	616	680	556	432	340
2.800	620	616	710	556	432	340
2.500	640	616	741	556	432	358
2.240	640	616	803	556	432	358
2.000	660	679	803	556	432	358
1.800	680	679	803	556	432	358
1.600	700	679	865	556	432	358
1.400	700	679	865	556	432	358
1.250	720	741	865	556	432	358
1.120	720	741	865	556	432	358
1.000	740	741	895	616	432	358
0.900	740	741	926	616	432	358
0.800	760	804	957	616	432	358
0.710	760	804	988	616	432	358
0.630	780	804	988	616	432	358
0.560	800	863	1,019	616	432	358
0.500	820	863	1,050	616	432	358
0.450	820	863	1,050	616	432	358
0.400	840	863	1,050	616	432	358
0.355	840	918	1,050	616	432	358
0.315	840	918	1,050	616	432	358
0.280	860	918	1,050	616	432	358
0.250	860	918	1,050	616	432	358
Torsional modulus	68.95×10^3	82.74×10^3	82.74×10^3	82.74×10^3	82.74×10^3	44.82×10^3
Bending modulus	200×10^3	200×10^3	200×10^3	200×10^3	200×10^3	12.41×10^3
Density	76.81×10^6	76.81×10^6	76.81×10^6	81.43×10^6	81.43×10^6	86.85×10^6

TABLE 8.9

Spring Wire Data: Variation of Material Strength with Temperature

Material Temperature	Stainless Steel S205	Steel	Steel 201 Def. 106	Nimonic 90 HR501	Nimonic 90 HR502	HD Phos. Brz. BS 384
20	1.000	1.000	1.000	1.000	1.000	1.000
100	0.980	0.958	0.958	1.000	1.000	1.000
150	0.966	0.933	0.933	1.000	1.000	0.000
200	0.955	0.917	0.917	1.000	1.000	0.000
225	0.000	0.000	0.000	0.907	1.000	0.000
250	0.000	0.000	0.000	0.857	0.997	0.000
275	0.000	0.000	0.000	0.857	0.987	0.000
300	0.000	0.000	0.000	0.857	0.975	0.000
325	0.000	0.000	0.000	0.000	0.958	0.000
350	0.000	0.000	0.000	0.000	0.940	0.000
375	0.000	0.000	0.000	0.000	0.913	0.000
400	0.000	0.000	0.000	0.000	0.887	0.000
425	0.000	0.000	0.000	0.000	0.855	0.000
450	0.000	0.000	0.000	0.000	0.822	0.000
475	0.000	0.000	0.000	0.000	0.783	0.000
500	0.000	0.000	0.000	0.000	0.733	0.000
525	0.000	0.000	0.000	0.000	0.688	0.000
550	0.000	0.000	0.000	0.000	0.633	0.000
575	0.000	0.000	0.000	0.000	0.583	0.000
600	0.000	0.000	0.000	0.000	0.517	0.000

TABLE 8.10

Spring Wire Data: Maximum Allowable Temperatures and Variations of Bending and Torsional Modulus with Temperature

Material Type	Maximum Temperature
S205	200°C
S201	200°C
DEF 106	200°C
Nimonic 90 HR501	300°C
Nimonic 90 HR502	600°C
BS 384 HD wire	100°C

Material Temperature	Stainless Steel S205	Steel S201	Steel Def. 106	Mimonic 90 HR501	Nimonic 90 HR502	ID Phos. Brz. BS 384
100	1.000	1.000	1.000	1.000	1.000	1.000
200	0.980	0.958	0.958	1.000	1.000	1.000
300	0.966	0.933	0.933	0.992	0.992	0.000
600	0.955	0.917	0.917	0.979	0.979	0.000

8.3 NOTES ON THE DESIGN OF HELICAL COMPRESSION SPRINGS MADE FROM ROUND WIRE

8.3.1 GENERAL NOTES

This section presents basic concepts and principles relating to the design and operation of helical compression springs manufactured from round wire.

Detailed treatment of their design is covered in Section 8.1, and the allowable stresses are covered in Section 8.2.

Springs having a constant wire diameter, coil diameter, and pitch are the essential concern of this chapter; deviations from these conditions are only permitted within the normal manufacturing tolerances.

There are many wide applications requiring compression springs, including:

- To store and return energy as in a cam-operated system
- To apply and maintain a constant force as in a relief valve
- To attenuate vibration as in a machine structure

The energy stored in a spring is proportional to the square of the stress in the spring. Therefore, for the most efficient use of the volume of the spring material, the highest reasonable stress levels should be used as the basis of design.

Springs used in the most critical of applications, such as a pressure relief valve, require the most careful attention to every possible detail of the spring performance and manufacture. The cost of the spring should be viewed in relation to the cost of the associated equipment and replacement if a failure occurred.

Not all springs perform such critical duties, and since extra precision adds to the part cost, careful attention needs to be given to the function of each spring and tolerances relaxed according to its application.

It needs to be stressed that all the information resulting from the design considerations is passed on to the spring manufacturer in a clear and concise manner. A checklist for the manufacturing requirements is covered in Section 8.3.7.6.

8.3.2 PRESTRESSING

In Section 8.2.4.1, the prestressing of the compression spring was considered. Prestressing (also known as scragging or presetting) induces beneficial residual stresses into the spring at the manufacturing stage. To carry out this process the spring is wound over length and is compressed to its solid length a number of times, usually five or six. This induces plastic deformation in the outer fibers of the spring coil. On release of all of the load, owing to the plastic flow that has occurred, the outer fibers of the coil, which in operation are the most highly stressed, are subjected to a residual shear stress by the inner fibers. This shear stress is in the opposite sense to that applied in the operation. The effect is to reduce the maximum stress, although the stress range remains unchanged. Therefore, the static strength and fatigue endurance of the spring are increased.

Figure 8.10 shows the spring load/deflection plot for a spring being prestressed. The first loading cycle is OPQR. Successive cycling of the load leads to a new load/deflection curve R_nQ parallel to OP, where R_n is the number of cycles the spring has been subject to.

The process also has the effect of increasing the apparent elastic limit of the material. The result of this prestressing means that in static loading the load carrying capacity of the spring is increased.

The improvements in the fatigue life and the static strength increase up to an optimum value with increasing overstrain. The amount of the increase and optimum overstrain will vary with the material. Discussions with the spring manufacturer will ensure the best possible results from the prestressing will be attained and the final details need not concern the designer.

8.3.3   CHOICE OF MATERIAL

The choice of material is dictated by the following considerations.

8.3.3.1   Operation Reliability

Material of high quality and assured properties are required if the spring is to operate for a large number of cycles (i.e., in excess of 10^7 cycles) or at an unavoidably high stress (e.g., because of space restrictions), or because a failure will be particularly serious. Alternatively, materials of a lower quality may be acceptable where the spring is only loaded infrequently or the load is constant or is lightly stressed, or if replacement of the spring is simple and acceptable. Where the spring weight is to be kept to a minimum (as in aerospace applications) alternative materials such as titanium alloy may be acceptable.

8.3.3.2   Corrosion and Protection

Corrosion (including stress corrosion) is a common problem with many spring materials, and as a result, their fatigue strength may be seriously reduced. Particular care must be exercised in selecting a suitable material and surface finish depending up on the environment the spring will be operating in.

Usually it is best practice to use a nonreactive material rather than to attempt to protect the material that is prone to corrosion. If the corrosion protection were to break down, then the spring would be subject to corrosion attack.

If it is necessary to provide a surface protection, then it may be achieved by various methods. Plating should be done to an approved specification (such as BS 1706), although in arduous conditions pin holes are always likely to occur in the plating, leading to a general breakdown in the corrosion protection.

Painting or coating with a plastic film is feasible, but again, failure of the protection is always likely at the loading points or if the coils contact.

Immersion in oil or grease may be a means of providing the required protection. Titanium alloy springs give a greatly improved service if the cost will allow, but will require anodizing after manufacture.

If shot peening is carried out the surface of the spring is in a chemically active condition and must be protected immediately after peening. Protection should be an invariable sequel to shot peening, but it is advisable to make this clear when ordering.

8.3.3.3   Working Temperature

At high temperatures, oxidation and general reduction of material properties, such as stress relaxation, make carbon steel springs generally unacceptable. In such cases it may be necessary to consider using alloy steels or suitable nonferrous materials.

At low subzero (cryogenic) temperatures allotropic or embrittlement may be a problem, and again it may be necessary to avoid carbon steels and some alloy steels.

A guide to the suitability of a number of acceptable spring materials is listed in Tables 8.11 and 8.12. For more specific information on the material properties, reference should be made to the quoted standard or to the manufacturer.

8.3.3.4   Special Requirements

The choice of material may be dictated by specific physical properties that may be required in an application. As an example, instrument springs may require exceptional stability of the material properties or low hysteresis. A special alloy Ni-span C902 may be used when a constant modulus over a range of temperatures is required.

8.3.4   LOADING

Depending how the spring is loaded will affect the permissible stress level and the magnitude of the induced stress.

TABLE 8.11
Alloy Steel Springs

Material		Specification	Maximum Wire Diameter mm	Normal Working Temperature Range of Spring °C	
Type				Minimum	Maximum
Carbon steels		BS 5216: HS2	13.25	−20	+130
		BS 5216: HS3	13.25	−20	+130
		BS 970: Pt. 5: 070A72	12.0	−20	+150
		BS 970: Pt. 5: 070A78	12.0	−20	+150
		BS 2803: 094A65 HS	12.5	−20	+150
Low-alloy steels	Silicon manganese	BS 970: Pt. 5: 250A58	40.0	−20	+150
	0.6–0.9% chromium	BS 970: Pt. 5: 250A61	40.0	−20	+150
		BS 970: Pt. 5: 527A60	80.0	−20	+150
	0.5% nickel-chromium Molybdenum	BS 970: Pt. 5: 805A60	80.0	−40	+150
	1% chromium-vanadium	BS 970: Pt. 5: 735A50	40.0	−20	+175
Alloy steels	Austenitic chromium-nickel	BS 2056: EN58A	10.0	−200	+250
		BS 970: Pt. 4: 302S25	25.0	−200	+250
	Austenitic chromium-nickel-molybdenum	BS 2056: EN58J	10.0	−200	+250
		BS 970: Pt. 4: 416S16	25.0	−200	+250
	17-7 chromium-nickel precipitation hardening	DTD 5086[a]	10.0	−90	+320
		FV 520B[b]	70.0	−90	+400

[a] UK Ministry of Defense specification.
[b] UK Trade description.

TABLE 8.12
Nonferrous Materials

Material		Specification	Maximum Wire Diameter mm	Normal Working Temperature Range of Spring °C	
Type				Minimum	Maximum
Phosphor-bronze with 8% tin		BS 2873 / ASTM B159-46	6.5	0	+100
Beryllium-copper with 2% beryllium		BS 2873 / ASTM B197-51T	6.5	−50	+100
Ni-span C902[a]		—	3.0	−50	+100
Nickel copper (Monel K-400)[b]		BS 3075-NA13	10.0	−40	+200
Nickel-copper-aluminum (Monel K-500)[b]		BS 3075-NA18	8.0	−200	+230
Nickel-chromium-iron (Inconel 600)[b]		BS 3075-NA14	10.0	−200	+370
Nickel-chromium-cobalt (Nimonic 90)[b]		BS 3075-2HR	8.0	−200	+550
Nickel-chromium P.H. (Inconel X-750)[b]		—	12.5	−200	+540
Titanium alloy (Ti-6A1-4V)		IMI 318[b]			+125

[a] U.S. trade description.
[b] UK trade description.

It is essential to ensure that the spring is not stressed outside its elastic range during manufacture, assembly, or operation. The nominal working stress should be low enough to preclude overstressing. The additional loading factors described below should be taken into consideration when determining the nominal working stress, as these will tend to increase the stress values if not considered.

8.3.4.1   Cyclic Loading and Fatigue Properties

If the spring is subject to cyclical loading, consideration should be given to the fatigue life and endurance of the selected material. There are two factors that that will increase the working stress.

1. The nonuniform distribution of strain along the length of the spring
2. Natural frequency (see Section 8.1.12)

The actual size of these effects will depend upon the relationship of the loading frequency to the natural frequency of the spring. The natural frequency will only be known when the final dimensions of the spring are fixed.

8.3.4.2   Transverse Loading

When the spring is subject to a transverse load, this does not occur in isolation; it is generally combined with an axial load, as in a vibratory conveyor, for example.

The stress increase due to the transverse load will need to be determined to ensure that premature failure will not occur. The combined transverse and axial stress must not exceed the stress when the spring is compressed to its solid length due to the axial load alone. If there is a danger that this may occur, then the designer should provide a stop to prevent any excessive deflection occurring in both planes. The design procedure is covered in detail in Section 8.1.9. The designer should pay particular attention to the spring housing and end locations to ensure the spring remains in position under the influence of the transverse load.

8.3.4.3   Impact

In some designs the compression spring will be subject to a shock load (as in the case of an engine valve or pneumatic drill). The shock loading will create a surge wave of torsional stress along the wire of the spring. If the spring is subject to shock loading it will be necessary to reduce the design stress to minimize the spring being inadvertently overstressed.

8.3.4.4   Eccentric Loading

An eccentric load will have a significant effect on the spring behavior. The load/deflection characteristics will deviate from the axially loaded spring, and there will be a greater tendency to buckle where the maximum stresses will be higher.

A further problem with compression springs is that the eccentricity of the load in respect to the axis of the spring will induce uneven loading around each coil, causing the coil to warp, or take up a form other than the true helical shape. Helix warping will increase the maximum stress in the spring, but an increase in the number of coils will reduce the effects of the warping. Helix warping in compression springs is covered in more detail in Section 8.1.11.

8.3.4.5   Buckling

In Section 8.1.8 it was stated that buckling of an unrestrained compression spring will occur when the length/mean diameter ratio exceeds 4.

It is recommended that all spring designs be checked against buckling. The onset of buckling will be affected greatly by the type of seating, which may be represented by the theoretical end condition factor C. These factors are considered in Figure 8.2 for the various types of seating. It has been found that compression springs tend to buckle earlier than the predicted theory due to spring ends tending to lift off on a nonpivoted seating as the critical buckling deflection is approached; there may be some slight initial curvature of the spring axis and out-of-squareness of the ends.

If the spring as designed is found to be unstable and the dimensions cannot be altered to avoid this condition, it may be possible to restrict the lateral deflection of the spring by means of a close-fitting housing or rod. This may lead to fretting fatigue failure. When a long spring, say 10 to 20 times the mean diameter, is required, it may be preferable to stack several springs over a rod using spigotted washers between each spring, and this will reduce the tendency to buckle.

8.3.5   DESIGN FEATURES

8.3.5.1   End Forms

In a compression spring the center coils, remote from the ends, are the truly functional part of the spring; the shape and form of the end coils are extremely important. The end shape will govern the load carrying capacity. The type of end coil fabrication will affect both the free length of the unloaded spring and its solid length when fully compressed. Section 8.1.3 gives guidance on the estimation of the free length, and Section 8.1.7 covers the solid length calculations for each type of end coil formation.

It is good practice to locate the ends of the spring in either a recess or a spigot, particularly when there are significant lateral loads.

There are various forms used for the end coils of compression springs. The more common ones are those used in Figure 8.14:

a. Coils left unclosed and not ground.
b. Coils left unclosed, but ends are ground flat, square to the axis.
c. One or more coils at each end closed but not ground.
d. One or more coils at each end closed and ground flat, square to the axis.

It must be remembered that the more rigorous the requirements, the more expensive the spring will become. A spring of the form 8.14(d) could cost as much as three times the cost of 8.14(a).

Springs with end forms of the type 8.14(d) will be preferred for precision applications such as spring balances, etc. This form will give the minimum load tolerance and have the minimum tendency to buckle; the loading will be carried more uniformly through the spring. It will also fit in its

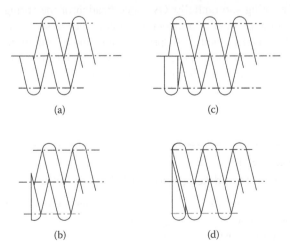

(a)   (c)

(b)   (d)

FIGURE 8.14   Spring end forms.

seat without deformation of the end coils and provide good axiality of the load. The spring rate will also be more uniform over the whole loading range.

Form 8.14(c) is cheaper to produce than form 8.14(d), and while the end will settle into its seat, an offset load is produced by the deformation of the end coils as the load is applied. This will induce early buckling.

Form 8.14(b) will be cheaper again; the end coils will deform more under load, which means the spring rate is not constant and the spring is much less stable.

Form 8.14(a) is in the as coiled state and is therefore the cheapest form to produce. Its application is very limited and is consequently only used where the application is not critical.

8.3.5.2 Free Length

The nominal free length of a compression spring is one of the basic dimensional requirements required by the spring manufacturer. It is the length of the spring in the unloaded state and is measured between two parallel plates perpendicular to the axis of the spring and in contact with the end coils, or part of the end coils (see Figure 8.14).

Normally a good quality compression spring is made over the length, so that after prestressing the material is elastic, and the free length is as specified by the designer.

8.3.5.3 Solid Length

The solid length of a compression spring is the overall dimension along the axis when all the coils are touching.

Compression springs are normally designed and manufactured so that the yield stress is not exceeded when they are compressed from the free length to the solid length. This is to avoid changes in the free length should the spring be accidentally compressed solid during its operation or assembly. The nominal solid lengths for this calculation are given in the Table 8.13 and are also necessary to avoid the possibility for some coils contacting at the spring's minimum working length due to irregular pitching. If some do touch, others may be overstressed and the spring rate may alter. Experience has shown the minimum clearance between coils should be 15% of the wire diameter, and this will normally avoid the problem.

Table 8.13 relates to springs manufactured from wire less than 13 mm in diameter.

8.3.5.4 Tolerances

Care should be exercised when assigning dimensions and tolerances. Unnecessarily close tolerances will raise the cost of the spring substantially. Overspecification of the spring dimensions and tolerances should be avoided; examples of this are not uncommon, such as specifying the outside diameter, wire diameter, and inside diameter together. A spring performing a particularly critical duty will have to have minimum tolerances in order to achieve the required precision and consistency in manufacture.

The following features will require tolerances to be defined:

Final load/length requirements
Spring rate

TABLE 8.13
Nominal Solid Length for End Coil Formations

End coil formation	Maximum solid length
One coil each end closed and ground flat	Nd
One coil each end closed but unground	$(N + 1.5)d$
End coil unclosed and unground	$(N + 1)d$

Wire diameter
Coil diameter (outside or inside)
Free length
Straightness
End squareness and parallelism

A suitable dimensioning and tolerancing system that helps to avoid these difficulties is given in BS 1726 Part 1 on helical compression springs.

It should be recognized that even with these acceptable dimensional and load tolerances, the resulting shear stress may vary by as much as ±12%.

8.3.5.5   Surface Finish

Surface defects on the wire (such as scratch-like grooves left by the drawing process) can lead to premature failure. It is the responsibility of the spring manufacturer to avoid these through good material specification and quality control. It is good design practice to draw attention to the surface finish requirements in the design specification.

It is also good practice to avoid identification marking, but if it is essential, the identification mark should be etched rather than stamped, as again this could lead to premature failure of the spring. The marking should be applied in an area of minimum stress, such as the end coils.

The use of ground wire will avoid any decarburization of the surface. If the compression spring is subsequently hot formed, the manufacturer will need to ensure that no further decarburization takes place during coiling or heat treatment.

8.3.5.6   Surface Treatment

As discussed in Section 8.2.4.2, shot peening is the process of bombarding the surface of the spring with small hard pellets at high velocity. This is carried out to improve the fatigue life of the spring by inducing beneficial residual stress in the outer surface of the wire. It may not be possible to carry this out on small-diameter or closely coiled springs, as accessibility to the inner surface will be difficult. In this instance, the result of shot peening is to leave the surface very susceptible to corrosion, and it must be protected immediately. Other surface treatments, including plating, painting, anodizing, and plastic coating, are acceptable methods of protecting the surface to reduce corrosion. None of these coating methods will affect the strength of the spring apart from improving the fatigue life of the spring.

8.3.6   Design Procedures

8.3.6.1   Basic Design

The physical dimensions of the compression spring may be determined from the design working stress, loading requirements, and space limitations. The recommended sequence for preparing a design is given in Section 8.1.2, followed by details of the design procedure.

8.3.7   Manufacturing Requirements

Unlike the manufacture of machine components, spring manufacturers require as much latitude as possible compatible with the real design requirements. To a certain extent, the spring duty may dictate what information the manufacturer requires. From a previous example, a pressure relief valve will require an accurate spring rate or an engine valve specifying two precise points for force and deflections.

In the case of specific design requirements such as the end fixing arrangements, fit over a spigot or inside a recess will need to be supplied in full.

TABLE 8.14

Example of Spring Manufacturing Requirements

Designation of Spring:				
Reference drawing number/part number:				
Material:				
Geometric data	Diameter (or gauge) of wire:		Tolerance	Only one diameter to be specified
	Inside coil diameter:		Tolerance	
	Mean diameter of coil:		Tolerance	
	Outside coil diameter:		Tolerance	
	Number of coils:	Working		
		Total		
	Type of ends:			
	Free length:			
	Solid length:			
Performance	Initial tension:		Tolerance	Give sufficient information without duplication of data
	Load P_1:		Tolerance	
	Load P_2:		Tolerance	
	Length at load P_1:		Tolerance	
	Length at load P_2:		Tolerance	
	Spring rate:		Tolerance	
Treatments	Heat treatment:			
	Prestressing:			State yes or no
	Surface treatment:			
	Surface protection:			
Details to be finalized	Maximum load:			
	Deflection for assembly:			
	Wire length:			
	Weight:			
	Quantity required:			
Other information				

The manufacturer will normally stock material to the standard wire gauge sizes over a wide range. In the case of compression springs manufactured from nonstandard wire diameter this may be obtained by grinding to size from standard wire before coiling. There may be a limit on the minimum diameter that can be obtained by this method, and a full discussion with the manufacturer will identify this.

It is essential that full descriptions of the material, geometric, surface finish, and end form requirements are provided if the spring is to meet the design requirements, with predictable and repeatable characteristics. Table 8.14 presents a suitable template showing the required spring details.

This data sheet is taken from BS 1726-1:2002.

It may not be possible for all the information to be identified due to the designer not always having the specialist knowledge and being able to prescribe all the requirements, but an experienced manufacturer will be able to supply this information, particularly for heat treatment, shot peening, and prestressing.

Table 8.15 should be completed with all the data required to manufacture in the spring. Table 8.16 will only be relevant to the designers and should be kept in the design file.

TABLE 8.15

Information Required by the Manufacturer

Designation of Spring:					
Reference drawing number/part number:					
Material:					
Geometric data	Diameter (or gauge) of wire:			Tolerance	Only one diameter to be specified
	Inside coil diameter:			Tolerance	
	Mean diameter of coil:			Tolerance	
	Outside coil diameter:			Tolerance	
	Number of coils:	Working			
		Total			
	Type of ends:				
	Free length:				
	Solid length:				
Performance	Initial tension:			Tolerance	Give sufficient information without duplication of data
	Load P_1:			Tolerance	
	Load P_2:			Tolerance	
	Length at load P_1:			Tolerance	
	Length at load P_2:			Tolerance	
	Spring rate:			Tolerance	
Treatments	Heat treatment:				
	Prestressing:				State yes or no
	Surface treatment:				
	Surface protection:				
Details to be finalized	Maximum load:				
	Deflection for assembly:				
	Wire length:				
	Weight:				
	Quantity required:				
Other information					

TABLE 8.16

Spring Specification Data Not Relevant to the Manufacturer

Shear stress in coils under load P_1:	
Shear stress in coils under load P_2:	
Shear stress in coils at solid length:	
Design life cycles:	
Maximum allowable temperature:	
Minimum allowable temperature:	
Environmental limitations:	
Natural frequency:	
Special notes:	

8.4   NESTED HELICAL COMPRESSION SPRINGS

8.4.1   General Notes to Section 8.4

When there is insufficient space available to fit a single spring without exceeding the maximum permissible stress, a design option may be to fit a nested spring, i.e., a spring within a spring.

The design of nested springs is treated here as a modification of the single spring, which satisfies the condition of loading, but not of space or stress.

The nest of two or three springs will give a higher volumetric efficiency, which may be used to either:

1. Reduce the outside diameter
2. Reduce the length
3. Reduce the maximum stress

Tables 8.18 to 8.20 enable a nest of springs to be derived from an equivalent single spring according to the requirements of either 1, 2, or 3.

Clearance between the coils is provided by a factor x, illustrated in Figure 8.15. Clearance between the nest and the housing requires a separate allowance. Recommended values of the clearance factor are given in Table 8.17.

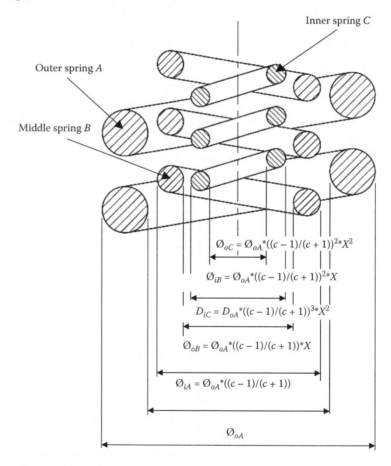

$$\varnothing_{oC} = \varnothing_{oA}*((c-1)/(c+1))^{2}*X^{2}$$

$$\varnothing_{iB} = \varnothing_{oA}*((c-1)/(c+1))^{2}*X$$

$$D_{iC} = D_{oA}*((c-1)/(c+1))^{3}*X^{2}$$

$$\varnothing_{oB} = \varnothing_{oA}*((c-1)/(c+1))*X$$

$$\varnothing_{iA} = \varnothing_{oA}*((c-1)/(c+1))$$

$$\varnothing_{oA}$$

FIGURE 8.15   Nest of three springs.

TABLE 8.17
Recommended Values of Clearance Factor x

Outside Diameter of Inner Spring	Clearance	Inside Diameter of Outer Spring	Value of x
20.0	0.063	20.5	0.92
25.0	0.079	25.0	0.93
50.0	0.128	50.0	0.94
75.0	0.159	80.0	0.95
100.0	0.179	105.0	0.95
125.0	0.208	130.0	0.95
150.0	0.228	160.0	0.96
180.0	0.258	180.0	0.97

Each spring in a nest should have the same effective solid length and the same deflection for any given stress. It follows that each spring must have the same index c. Then if D_o of the outer spring and c are known, the coil and wire diameters can be written as in Figure 8.15.

Outside diameter of nests of two or three springs with clearance factors of 0.98, 0.94, and 0.90. The diameters are stated as fractions of the outside diameter of the single spring designed for the same load, stress, and effective solid length.

TABLE 8.18

To Reduce the Outside Diameter

	Nn = 2			Nn = 3		
	Clearance Factors					
Index c	**0.98**	**0.94**	**0.90**	**0.98**	**0.94**	**0.90**
2.0	0.951	0.954	0.958	0.944	0.950	0.955
3.0	0.898	0.905	0.912	0.878	0.888	0.897
4.0	0.862	0.871	0.880	0.826	0.839	0.852
4.1	0.859	0.869	0.877	0.822	0.836	0.849
4.2	0.856	0.866	0.875	0.818	0.832	0.845
4.3	0.853	0.863	0.872	0.814	0.828	0.841
4.4	0.851	0.861	0.870	0.810	0.824	0.838
4.5	0.848	0.858	0.868	0.806	0.821	0.835
4.6	0.846	0.856	0.865	0.802	0.817	0.832
4.7	0.844	0.854	0.863	0.798	0.814	0.829
4.8	0.842	0.852	0.861	0.796	0.810	0.825
4.9	0.840	0.850	0.859	0.792	0.807	0.822
5.0	0.837	0.847	0.858	0.788	0.804	0.819
5.1	0.835	0.845	0.856	0.785	0.801	0.816
5.2	0.833	0.843	0.854	0.782	0.798	0.814
5.3	0.831	0.841	0.853	0.779	0.795	0.811
5.4	0.829	0.840	0.851	0.776	0.792	0.808
5.5	0.828	0.838	0.849	0.773	0.790	0.806
5.6	0.826	0.837	0.848	0.770	0.787	0.804
5.7	0.824	0.835	0.846	0.768	0.785	0.801
5.8	0.822	0.834	0.844	0.765	0.782	0.799
5.9	0.820	0.832	0.843	0.763	0.780	0.797
6.0	0.819	0.830	0.841	0.760	0.777	0.795
6.1	0.817	0.829	0.840	0.758	0.775	0.793
6.2	0.816	0.827	0.838	0.756	0.773	0.791
6.3	0.815	0.826	0.837	0.753	0.771	0.789
6.4	0.814	0.825	0.836	0.751	0.769	0.757
6.5	0.812	0.824	0.835	0.749	0.767	0.785
6.6	0.811	0.822	0.833	0.747	0.765	0.783
6.7	0.810	0.821	0.832	0.745	0.763	0.781
6.8	0.808	0.820	0.831	0.743	0.761	0.779
6.9	0.807	0.819	0.830	0.741	0.759	0.777
7.0	0.806	0.817	0.829	0.739	0.757	0.775
7.1	0.805	0.816	0.828	0.737	0.756	0.774
7.2	0.804	0.815	0.827	0.735	0.754	0.772
7.3	0.802	0.814	0.826	0.734	0.752	0.770
7.4	0.801	0.813	0.825	0.732	0.751	0.769

TABLE 8.18 *(Continued)*

To Reduce the Outside Diameter

	Nn = 2			Nn = 3		
			Clearance Factors			
Index c	0.98	0.94	0.90	0.98	0.94	0.90
7.5	0.800	0.812	0.824	0.730	0.749	0.768
7.6	0.799	0.811	0.823	0.728	0.747	0.766
7.7	0.798	0.810	0.822	0.726	0.746	0.765
7.8	0.797	0.809	0.821	0.725	0.744	0.763
7.9	0.796	0.808	0.820	0.723	0.742	0.762
8.0	0.795	0.807	0.819	0.722	0.741	0.761
8.1	0.795	0.806	0.818	0.721	0.740	0.759
8.2	0.794	0.805	0.818	0.719	0.738	0.758
8.3	0.793	0.804	0.817	0.718	0.737	0.757
8.4	0.792	0.804	0.816	0.716	0.736	0.755
8.5	0.791	0.803	0.815	0.715	0.735	0.754
8.6	0.790	0.802	0.814	0.714	0.734	0.753
8.7	0.789	0.801	0.814	0.712	0.733	0.752
8.8	0.789	0.801	0.813	0.711	0.732	0.751
8.9	0.788	0.800	0.812	0.710	0.730	0.749
9.0	0.787	0.799	0.811	0.708	0.827	0.748
10.0	0.780	0.793	0.805	0.698	0.718	0.738

TABLE 8.19

To Reduce Stress

	Nn = 2			Nn = 3		
			Clearance Factors			
Index of Single Spring	0.98	0.94	0.90	0.98	0.94	0.90
2.0	2.05	2.04	2.04	2.06	2.05	2.04
3.0	3.16	3.15	3.13	3.20	3.18	3.16
4.0	4.28	4.26	4.24	4.38	4.34	4.31
4.1	4.40	4.37	4.35	4.50	4.46	4.43
4.2	4.51	4.48	4.46	4.62	4.58	4.55
4.3	4.62	4.59	4.57	4.74	4.70	4.66
4.4	4.74	4.71	4.69	4.86	4.82	4.78
4.5	4.85	4.83	4.80	4.98	4.94	4.90
4.6	4.97	4.94	4.91	5.10	5.06	5.02
4.7	5.08	5.05	5.02	5.28	5.18	5.13
4.8	5.19	5.17	5.14	5.34	5.30	5.25
4.9	5.31	5.28	5.25	5.46	5.42	5.37
5.0	5.42	5.39	5.36	5.58	5.53	5.48
5.1	5.53	5.50	5.47	5.70	5.65	5.60

TABLE 8.19 *(Continued)*
To Reduce Stress

Index of Single Spring	Nn = 2			Nn = 3		
	Clearance Factors					
	0.98	0.94	0.90	0.98	0.94	0.90
5.2	5.64	5.61	5.58	5.82	5.77	5.71
5.3	5.75	5.72	5.69	5.95	5.89	5.83
5.4	5.87	5.83	5.80	6.07	6.01	5.95
5.5	5.98	5.95	5.91	6.19	6.13	6.07
5.6	6.10	6.06	6.03	6.32	6.25	6.19
5.7	6.22	6.18	6.15	6.45	6.38	6.31
5.8	6.33	6.30	6.26	6.57	6.50	6.43
5.9	6.44	6.41	6.37	6.69	6.62	6.54
6.0	6.56	6.52	6.48	6.81	6.76	6.66
6.1	6.67	6.63	6.60	6.93	6.86	6.78
6.2	6.78	6.74	6.71	7.05	6.98	6.90
6.3	6.90	6.86	6.82	7.17	7.10	7.02
6.4	7.02	6.97	6.93	7.30	7.22	7.14
6.5	7.13	7.08	7.04	7.42	7.34	7.26
6.6	7.24	7.20	7.15	7.54	7.46	7.38
6.7	7.36	7.31	7.27	7.66	7.57	7.49
6.8	7.47	7.42	7.38	7.78	7.69	7.61
6.9	7.58	7.53	7.49	7.90	7.81	7.73
7.0	7.69	7.64	7.60	8.03	7.93	7.84
7.1	7.81	7.76	7.71	8.15	8.05	7.96
7.2	7.92	7.87	7.82	8.27	8.17	8.08
7.3	8.04	7.99	7.93	3.39	8.29	8.20
7.4	8.15	8.10	8.04	8.52	8.41	8.32
7.5	8.27	8.22	8.16	8.65	8.54	8.44
7.6	8.38	8.34	8.28	8.77	8.66	8.56
7.7	8.49	8.45	8.39	8.89	8.78	8.68
7.8	8.61	8.56	8.50	9.01	8.90	8.79
7.9	8.72	8.67	8.61	9.13	9.02	8.91
8.0	8.84	8.78	8.72	9.25	9.14	9.03
8.1	8.95	8.89	8.84	9.38	9.26	9.15
8.2	9.06	9.00	8.95	9.50	9.38	9.27
8.3	9.18	9.12	9.06	9.62	9.50	9.39
8.4	9.29	9.23	9.17	9.75	9.63	9.51
8.5	9.41	9.34	9.28	9.87	9.75	9.63
8.6	9.52	9.45	9.39	9.99	9.87	9.75
8.7	9.64	9.57	9.51	10.11	9.99	9.87
8.8	9.75	9.68	9.62	10.23	10.11	9.99
8.9	9.87	9.8	9.73	10.36	10.23	10.10
9.0	9.98	9.91	9.84	10.48	10.35	10.22
10.0	11.12	11.04	10.98	11.71	11.56	11.42

TABLE 8.20

To Reduce Length

Index of	Nn = 2			Nn = 3		
	Clearance Factors					
Single Spring	0.98	0.94	0.90	0.98	0.94	0.90
2.0	2.18	2.16	2.14	2.23	2.20	2.17
3.0	3.40	3.37	3.34	3.52	3.47	3.42
4.0	4.64	4.59	4.54	4.90	4.81	4.72
4.1	4.76	4.71	4.66	5.04	4.94	4.85
4.2	4.88	4.83	4.78	5.18	5.08	4.99
4.3	5.01	4.95	4.9	5.32	5.22	5.12
4.4	5.13	5.08	5.02	5.46	5.35	5.25
4.5	5.26	5.20	5.14	5.6	5.48	5.38
4.6	5.38	5.32	5.26	5.74	5.62	5.51
4.7	5.50	5.44	5.38	5.88	5.76	5.65
4.8	5.63	5.57	5.51	6.02	5.90	5.79
4.9	5.76	5.69	5.63	6.16	6.04	5.92
5.0	5.88	5.81	5.75	6.30	6.17	6.04
5.1	6.00	5.94	5.87	6.44	6.30	6.18
5.2	6.13	6.06	5.99	6.59	6.44	6.31
5.3	6.25	6.18	6.11	6.73	6.58	6.44
5.4	6.38	6.31	6.24	6.87	6.72	6.58
5.5	6.50	6.43	6.36	7.01	6.86	6.71
5.6	6.63	6.55	6.48	7.15	7.00	6.85
5.7	6.76	6.68	6.60	7.29	7.13	6.98
5.8	6.80	6.80	6.72	7.43	7.27	7.11
5.9	7.01	6.92	6.84	7.57	7.40	7.25
6.0	7.13	7.04	6.96	7.72	7.54	7.38
6.1	7.26	7.17	7.08	7.86	7.68	7.72
6.2	7.38	7.30	7.20	8.00	7.82	7.65
6.3	7.51	7.42	7.33	8.14	7.96	7.78
6.4	7.64	7.54	7.45	8.28	8.10	7.92
6.5	7.70	7.67	7.57	8.42	8.25	8.05
6.6	7.89	7.79	7.69	8.56	8.27	8.18
6.7	8.01	7.91	7.81	8.71	8.50	8.32
6.8	8.14	8.03	7.94	8.85	8.64	8.45
6.9	8.26	8.16	8.06	8.99	8.77	8.59
7.0	8.38	8.28	8.18	9.13	8.91	8.72
7.1	8.51	8.40	8.30	9.27	9.05	8.85
7.2	8.63	8.52	8.43	9.42	9.19	8.99
7.3	8.76	8.65	8.55	9.56	9.33	9.12
7.4	8.88	8.77	8.67	9.70	9.47	9.26

TABLE 8.20 *(Continued)*
To Reduce Length

Index of Single Spring	Nn = 2			Nn = 3		
	Clearance Factors					
	0.98	0.94	0.90	0.98	0.94	0.90
7.5	9.00	8.89	8.79	9.84	9.61	9.39
7.6	9.12	9.01	8.91	9.98	9.75	9.53
7.7	9.25	9.14	9.03	10.12	9.89	9.66
7.8	9.37	9.25	9.15	10.27	10.03	9.80
7.9	9.50	9.38	9.27	10.41	10.16	9.93
8.0	9.63	9.51	9.39	10.55	10.30	10.07
8.1	9.75	9.63	9.51	10.69	10.44	10.20
8.2	9.88	9.75	9.64	10.83	10.58	10.34
8.3	10.00	9.88	9.76	10.97	10.72	10.47
8.4	10.13	10.00	9.88	11.12	10.86	10.61
8.5	10.25	10.12	10.00	11.26	10.99	10.74
8.6	10.38	10.25	10.13	11.40	11.13	10.88
8.7	10.50	10.37	10.25	11.54	11.27	11.01
8.8	10.63	10.52	10.37	11.68	11.41	11.15
8.9	10.75	10.62	10.49	11.82	11.55	11.28
9.0	10.88	10.74	10.61	11.97	11.68	11.41
10.0	12.14	11.98	11.83	13.40	13.06	12.76

Values of index C for nests of two and three springs with clearance factors of 0.98, 0.94, and 0.90. The values of the index are tabulated against the index of the single spring, which has the same outside diameter, load, deflection, and effective solid length. The stress in each spring of the nest designed from this table will be lower than the stress in the corresponding single spring.

Values of the index c for nests of two or three springs with clearance factors of 0.98, 0.94, and 0.90. The values of the index are tabulated against the index of the single spring, which has the same outside diameter, load, deflection, and stress.

The length of each spring of a nest designed from this table will be less than the corresponding single spring.

8.4.2 EXAMPLE 2

The following example considers replacing a single spring with a nest of three springs.

The single spring has the following characteristics:

D_o = 120.0 mm
d = 20.0 mm
D = 100.0 mm
C = 5.0
n = 8
L_{cl} = n.d = 160 mm

L_c = $(n - 1)d = 180$
P = 62,560 N
δ = 86.36 mm
S = 201.4 N/mm
K = 1.3
q = 696.4 MPa

Let L_{cl} and q be retained. Find the outside diameter of a nest of three springs.

Section 8.1.5.2 shows that the maximum stress q remains unchanged only if c = 5.0, and therefore the stress concentration factor K is retained.

As the inside diameter of the coil of the single spring is $D_i = (D - d) = 80.0$ mm, the clearance factor recommended in Table 8.17 is x = 0.95.

From Table 8.18 for Nn = 3, x = 0.95 and c = 5.0, the ratio D_o (nest)/D_o (single spring) = 0.79 is found by interpolation. Hence D_o (nest) = 0.79 × 120.0 = 94.8 mm.

The outside diameters of the other two springs will be obtained by means of the terms given in Figure 8.15.

Namely:

$$\left(\frac{c-1}{c+1}\right) x = \frac{4}{6} .0.95 = 0.6333$$

and

$$\left(\frac{c-1}{c+1}\right)^2 \cdot x^2 = 0.401$$

The outside diameter of the middle and inner spring is obtained from Figure 8.15:

$$D_{o\ middle} = D_o \cdot \left(\frac{c-1}{c+1}\right) \cdot x$$

$$= 94.8 \cdot \left(\frac{5-1}{5+1}\right) \cdot 0.95$$

$$= 60.04 \text{ mm.}$$

$$D_{o\ inner} = D_o \cdot \left(\frac{c-1}{c+1}\right)^2 \cdot x^2$$

$$= 94.8 \cdot \left(\frac{5-1}{5+1}\right)^2 \cdot 0.95^2$$

$$= 38.024 \text{ mm.}$$

Table 8.21 can now be constructed.

The reduction in the outside diameter (D_o) compared with the single spring:

It will also be noted that the solid length (Lc_l) is also shorter than the single spring.

This example of a spring nest is impracticable, as the inner spring is laterally unstable (see Section 8.3.4.5). The solution is to redesign the nest using only two springs. In this design the inner spring is stable and practicable.

TABLE 8.21
Summary of Nest Details for Example 2

D_o mm	94.8	60.04	38.03
$d = \dfrac{D_o}{c+1}$ mm	15.8	10.07	6.338
$D = D_o - d$ mm	79.0	49.97	31.69
$D_1 = D - 2d$ mm	63.2	39.90	25.35
$n = \dfrac{L_{c1}}{d}$ mm	10.13	15.89	25.24
$L_c = L_{c1} + d$ mm	175.8	170.07	166.34
$S = \dfrac{Dd^4}{8nD^3}$ N/mm	118.54	49.27	19.08 $\Sigma S = 186.89$ N/mm
$L_c = L_{c1} + d$ mm	175.85	170.08	166.31

TABLE 8.22
Summary of Nest Details for Example 2

D_o mm	100.74	63.80
$d = \dfrac{D_o}{c+1}$ mm	16.79	10.63
$D = D_o - d$ mm	83.95	53.17
$D_1 = D - 2d$ mm	67.16	42.54
$n = \dfrac{L_{c1}}{d}$ mm	9.529	15.05
$L_c = L_{c1} + d$ mm	176.79	170.63
$S = \dfrac{Dd^4}{8nD^3}$ N/mm	133.91	53.70 $\Sigma S = 187.60$ N/mm

From Table 8.22, considering N = 2, x = 0.95, and c = 5.0 and interpolating:

$$D_o = 0.8395 \times 120.0 \text{ mm}$$

$$= 100.74 \text{ mm}$$

The reduction in the outside diameter:

$$\text{Reduction} = \left(1 - \frac{100.74}{120.00}\right) \cdot 100$$

$$= 16.05\%$$

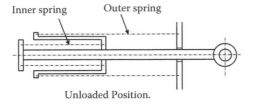

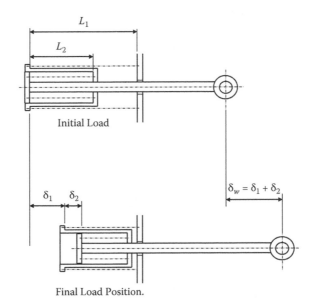

FIGURE 8.16   Nested springs in series.

8.4.3   NESTED SPRINGS IN SERIES

This arrangement of spring assembly (see Figure 8.16) is generally used to vertically balance a mass such as a cover. It has the advantage that the deflection can be large but the solid length is small. In a housing of specific dimensions, the spring assembly will have a lower stress and also offer a greater resistance to buckling compared to that of a single spring.

The proportions in which the total working deflection (δ_w) comprises the deflections δ_1 and δ_2.

These deflections will depend upon the space required to accommodate the tube that transmits the load from the inner spring to the outer spring.

An approximate estimate of these deflections can be determined from the following formula:

$$\frac{d_1}{7L_1} = \frac{d_2}{3L_2} = \frac{d_w}{7L_1 + 3L_2}$$

For example if $L_2 = 0.5\,L_1$, the proportion of the total deflection that is due to the outer spring will be:

$$\frac{d_1}{d_w} = \frac{7L_1}{7L_1 + 3L_2} = \frac{14L_2}{17L_2}$$

$$= 0.824$$

8.4.4   EXAMPLE 3

A telescopic spring is required to balance an assembly to meet the following requirements:

Load range   6,900–12,200 N
Corresponding working deflection (δ_w)   320 mm
Maximum working stress (q)   827.4 MPa
Stress to closure limited to   956.3 MPa
Outside diameter of the nest not to exceed   140.0 mm

The inner spring is to be half the length of the outer spring in the initial loaded condition.

Considering the outer spring:

$$L_1 = 2L_2 \quad \text{and} \quad \frac{\delta_1}{\delta_w} = \frac{14}{17}$$

$$= 0.824$$

$$\delta 1 = 0.824\,\delta_w$$

$$= 0.824 \cdot 320 \text{ mm.}$$

$$= 263.53 \text{ mm.}$$

Spring rate:

$$= \frac{12200 - 6900}{263.53}$$

$$= 20.11 \text{ N/mm}$$

Total deflection:

$$d = \frac{12200 \text{ N}}{20.11 \text{ N/mm}}$$

$$= 606.663 \text{ mm}$$

Initial deflection:

$$\delta_1 = \frac{6900 \text{ N}}{20.11 \text{ N/mm}}$$

$$C_2 = D_o \sqrt{\left(\frac{q}{P}\right)}$$

$$= 140.0 \sqrt{\left(\frac{827.3 \text{ MPa}}{12200 \text{ N}}\right)}$$

$$= 36.457$$

Hence, $c = 6.922$.

$\quad C_3 = 0.03974$ (by interpolation from Table 8.4).

Now:

$$C_3 = \sqrt{\left(\frac{d}{L_{c1}.q}\right)}$$

$$0.03974 = \sqrt{\left(\frac{606.663}{L_{c1} \cdot 827.37}\right)}$$

$$\therefore\ L_{c1} = 464.29\ \text{mm}$$

Deflection to closure:

$$\delta_c = 606.66\ \text{mm} \cdot \frac{956.27\ \text{MPa}}{827.32\ \text{MPa}}$$

$$= 701.22\ \text{mm}$$

Free length L_o':

$$= 701.22\ \text{mm} + 464 \cdot 3\ \text{mm}$$

$$= 1165.50\ \text{mm}$$

Length at initial load:

$$= L' = L_o' - \delta_t$$

$$= 1165.5\ \text{mm} - 343.113\ \text{mm}$$

$$= 822.387\ \text{mm}$$

Number of working coils:

$$n = \frac{L_{c1}}{d}$$

$$= \frac{464.3\ \text{mm}}{17.67\ \text{mm}}$$

$$- 26.27\ \text{coils}$$

Free length (L_o):

$$L_o = L_{c1} + 2d$$

$$= 1165.5\ \text{mm} + (2 \cdot 17.67\ \text{mm})$$

$$= 1200.844\ \text{mm}$$

Inner spring:

$$d_2 = d_w - d_1$$

$$d_2 = 320.0 \text{ mm} - 263.53 \text{ mm}$$

$$= 56.47 \text{ mm}$$

Spring rate:

$$= \frac{12200 \text{ N} - 6900 \text{ N}}{56.47 \text{ mm.}}$$

$$= 93.86 \text{ N/mm.}$$

Length of spring at initial load L_2:

$$L_2 = 0.5 \, L_1$$

$$= 0.5 \cdot 822.37 \text{ mm}$$

$$= 411.19 \text{ mm}$$

Initial deflection d_i:

$$d_i = \frac{6900 \text{ N}}{93.855 \text{ N/mm}}$$

$$= 73.52 \text{ mm}$$

Deflection to closure:

$$\delta_c = \frac{12200 \text{ N}}{93.855 \text{ N/mm}} \cdot \frac{956.27 \text{ MPa}}{827.32 \text{ MPa}}$$

$$= 150.25 \text{ mm.}$$

$$\therefore \ L_{c1} = L_2 + \delta_i - \delta_c$$

$$= 411.19 \text{ mm} + 73.52 \text{ mm} - 150.25 \text{ mm}$$

$$= 334.46 \text{ mm}$$

$$C_3 = \sqrt{\left(\frac{\delta}{L_{c1} \cdot q} \right)}$$

$$= \sqrt{\left(\frac{56.47 \text{ mm} + 73.52 \text{ mm}}{334.464 \text{ mm} \cdot 827.37 \text{ MPa}} \right)}$$

$$= 0.02167$$

By interpolation from Table 8.4:

$$c = 4.0617$$

$$C_2 = 19.2085$$

$$D_o = C_2 \sqrt{\left(\frac{P}{q}\right)}$$

$$= 19.2085 \sqrt{\left(\frac{12200 \text{ N}}{827.37 \text{ N/mm}^2}\right)}$$

$$= 73.76 \text{ mm}$$

Number of working coils n:

$$n = \frac{L_{cl}}{d}$$

$$= \frac{334.464 \text{ mm}}{14.572 \text{ mm}}$$

$$= 22.95 \text{ coils}$$

FURTHER READING

Ministry of Supply. Notes on the design of helical compression springs. In *Spring Design Memorandum 1*. London: HMSO, June 1942.

Wahl, A.M. Diametral expansion of helical compression springs during deflection. *J. Appl. Mech.*, 20, 1953.

9 Introduction to Analytical Stress Analysis and the Use of the Mohr Circle

9.1 INTRODUCTION

Of all the graphical methods used by an engineer there is little doubt that the Mohr circle is the best known; in its various applications it is an aid to visualizing a stress or strain problem.

The object of this chapter is to explore with a reasonable thoroughness three important uses of the circle. These applications confront the engineer and designer in many branches of mechanical and structural engineering.

In the development of the circle construction it has been considered desirable to present a concise account of the underlying theory. It is possible for the reader more interested in applications to accept these results and go straight to those sections dealing with the constructions and applications.

Nearly all textbooks, of necessity, deal with the circle in too cursory a manner and leave many detailed questions unanswered. It is hoped that this chapter devoted solely to the circle will answer more of these questions. In illustrating the text with examples of a practical nature other topics of strength of materials will be quoted, the background of which can be found in most strength of materials textbooks.

9.2 NOTATION

A = Area
a = Circle constant
B = Breadth (of beam cross section)
D = Depth (of beam section)
E = Modulus of elasticity
e = Strain (usually with suffix to indicate direction)
F = Force
h = Distance
I = Second moment of area (with suffix of the type xx or cg)
J = Polar moment (with suffix of the type xy)
M = Moment (with suffix to indicate axis)
n = Factor of safety
O − Origin of graph
P = Force, or pole point
$P_{1,2}$ = Pole points
R = Radius of circle, radius of curvature
T = Torque (usually with suffix to indicate axis)
y = Distance from neutral axis of a beam to a given point
β = Angle
γ = Poisson's ratio, angle

δ = Deflection
θ = Angle of plane
θ_p = Angle of principal plane
σ = Stress (with suffix of the type x to indicate plane)
σ_y = Yield stress, or normal stress in y direction
σ_w = Working stress
τ = Shear stress (with suffix of the type xy)
φ = Shear strain (with suffix of the type xy)

9.3   TWO-DIMENSIONAL STRESS ANALYSIS

Consider the rectangle ABCD shown in Figure 9.1 of unit thickness and subject to direct stresses, σ_x and σ_y, together with shearing stresses, τ_{xy}.

The stress normal to the plane of the diagram σ_z is equal to zero. This is the condition of plane stress.

It is necessary to investigate the stress conditions on any plane making an angle θ with the side AB and the consequence that follows from the existence of such stresses.

A triangular element is isolated from the rectangle, the stresses on the original planes AB and AE being known and the sloping side being an unknown normal stress of σ_n and an unknown shearing force τ are postulated. The problem is to obtain expressions for these unknown stresses by considering forces (not stresses, please note) acting on this triangular element of unit thickness. The length BE is assumed to be one unit as shown in Figure 9.2. The forces are then resolved in normal and tangential directions to the slope BE.

For equilibrium the sum of all the forces in a given direction must equal zero.

Thus resolving normally to BE in accordance with the sign convention accompanying Figure 9.2,

$$\Sigma P_n = -\sigma_x \, Cos\theta Cos\theta + \tau_{xy} \, Cos\theta Sin\theta + \tau_{xy} \, Cos\theta Sin\theta - \sigma_y \, Sin\theta Sin\theta + \sigma_n = 0$$

$$\sigma_n = \sigma_x \, Cos^2\theta + \sigma_y \, Sin^2\theta - 2\tau_{xy} \, Sin\theta Cos\theta$$

and in terms of double angles:

$$\sigma_n = \tfrac{1}{2}(\sigma x + \sigma y) + \tfrac{1}{2}(\sigma x - \sigma y) \, Cos2\theta - \tau_{xy} \, Sin2\theta \qquad (9.1)$$

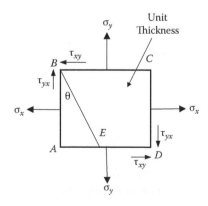

FIGURE 9.1   A unit thickness element subject to a biaxial stress field.

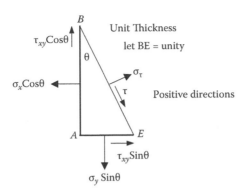

FIGURE 9.2 Section of element in Figure 9.1.

Resolving tangentially along BE and again carefully following the sign convention,

$$\Sigma P_\theta = -\sigma_x \, Cos\theta Sin\theta - \tau_{xy} \, Cos^2\theta + \sigma_y \, Sin\theta Cos\theta + \tau_{xy} \, Sin^2\theta + \tau = 0$$

$$\tau = \tfrac{1}{2}(\sigma_x - \sigma_y) \, Sin2\theta + \tau_{xy} \, Cos2\theta \tag{9.2}$$

9.4 PRINCIPAL STRESSES AND PRINCIPAL PLANES

σ_n is clearly a quantity that varies with the angle θ. By differentiating with respect to θ and equating to zero, the conditions for maximum and minimum normal stress can be found from Equation (9.1):

$$\frac{d\sigma_n}{d\theta} = -\left(\sigma_x - \sigma_y\right) Sin2\theta - 2\tau_{xy} Cos2\theta = 0 \tag{9.3}$$

Thus

$$Tan\, 2\theta = -\frac{2\tau_{xy}}{\left(\sigma_x - \sigma_y\right)} \tag{9.4}$$

A comparison of Equations (9.2) and (9.3) shows that τ is zero on the planes where the maximum or minimum normal stresses are found.

This is an extremely important finding: the conditions for σ_{max} or σ_{min} have been derived, and on the planes where these stresses occur the shear stresses are zero. There are two of these planes because there are two values of θ, 90° apart, which satisfy Equation (9.4).

These planes are called *principal planes*, and the normal stresses σ_{max} and σ_{min} acting on these planes are called *principal stresses*. On all the other planes shear stresses will be acting. It is also possible to show that in a three-dimensional system there are three mutually perpendicular planes that possess no shear stresses.

Figure 9.3(a) and (b) shows, respectively, the complex stress situation and the equivalent principal stress pattern. It must be clearly understood that these are equivalent systems. In the solution of a problem the engineer chooses the system that best fits the purpose. Generally the stresses shown in Figure 9.3(a) are obtained by some preliminary calculations and the principal stress system is then deduced. Frequently this second system is the one more easily assessed. Expressions for the principal stresses, now to be called σ_1 and σ_2, can be obtained from Equations (9.1) and (9.3), giving

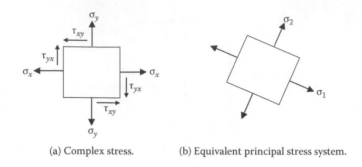

(a) Complex stress.   (b) Equivalent principal stress system.

FIGURE 9.3 Principal stresses and planes: complex stress system.

$$\sigma_{1,2} = 0.5(\sigma_x + \sigma_y) + \frac{0.5(\sigma_x - \sigma_y) \cdot 0.5(\sigma_x - \sigma_y) + \tau_{xy}{}^2}{\sqrt{\left(\dfrac{\sigma_x - \sigma_y}{2}\right)^2 + \tau_{xy}{}^2}}$$

$$\sigma_{1,2} = 0.5(\sigma_x + \sigma_y) \pm \sqrt{\left(\frac{\sigma_x - \sigma_y}{2}\right)^2 + \tau_{xy}{}^2} \tag{9.5}$$

9.4.1 MAXIMUM SHEAR STRESS

If now Equation (9.2) is differentiated and equated to zero, the condition for maximum shear stress within the element can be obtained:

$$\frac{d\tau}{d\theta} = (\sigma_x - \sigma_y)\mathrm{Cos}2\theta - 2\tau_{xy}\,\mathrm{Sin}2\theta = 0$$

Hence,

$$\mathrm{Tan}2\theta = \left(\frac{\sigma_x - \sigma_y}{2 \cdot \tau_{xy}}\right) \tag{9.6}$$

This is the negative reciprocal of Equation (9.3). Consideration of Equations (9.3) and (9.5) shows the values of 2θ defined by each differ by $90°$. Thus θ differs by $45°$ in each case. This means that the maximum shear stress occurs on planes $45°$ to the planes of principal stress. This is shown in Figure 9.4(a) and (b). It is worth noting that the maximum shears are accompanied by normal stresses.

Equation (9.6) can be used to obtain τ_{max} from Equation (9.2):

$$\tau_{max} = \frac{0.5(\sigma_x - \sigma_y)0.5(\sigma_x - \sigma_y) + \tau_{xy}{}^2}{\sqrt{\left(\dfrac{\sigma_x - \sigma_y}{2}\right)^2 + \tau_{xy}{}^2}} \tag{9.7}$$

$$= \pm\sqrt{\left(\frac{\sigma_x - \sigma_y}{2}\right)^2 + \tau_{xy^2}}$$

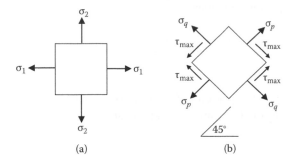

FIGURE 9.4   Principal stresses and planes: maximum shear stresses.

Expressions (9.5) and (9.7) lead to two simple and important relations:

i.e.,

$$\tau_{max} = \text{half the principal stress difference}$$

and

$$\sigma_1 + \sigma_2 = \sigma_x + \sigma_y = \text{a constant value}$$

Thus the sum of the normal stresses on any pair of perpendicular faces is invariant.

$$\tau_{max} = \frac{1}{2}(\sigma_1 - \sigma_2)$$

This small number of equations forms the basis of stress analysis. In the following sections the geometric interpretation of these equations will be considered.

9.4.2   GEOMETRIC INTERPRETATION

For the present purpose Equations (9.1) and (9.2) are rearranged as follows:

$$\sigma_n - \frac{1}{2}(\sigma_x - \sigma_y) = \frac{1}{2}(\sigma_x - \sigma_y) \cdot \text{Cos}2\theta - \tau_{xy} \cdot \text{Sin}2\theta$$

$$\tau = \frac{1}{2}(\sigma_x - \sigma_y) \cdot \text{Sin}2\theta + \tau_{xy} \cdot \text{Cos}2\theta$$

Squaring each side these become:

$$\left[\sigma_n - \frac{1}{2}(\sigma_x - \sigma_y)\right]^2 = \frac{1}{4}(\sigma_x - \sigma_y)^2 \cdot \text{Cos}^2 2\theta - (\sigma_x - \sigma_y) \cdot \tau_{xy} \cdot \text{Sin}2\theta \cdot \text{Cos}2\theta + \tau_{xy}{}^2 \text{Sin}^2 2\theta$$

$$\tau^2 = \frac{1}{4}(\sigma_x - \sigma_y)^2 \text{Sin}^2 2\theta + (\sigma_x - \sigma_y) \cdot \tau_{xy} \cdot \text{Sin}2\theta \cdot \text{Cos}2\theta + \tau_{xy}{}^2 \text{Cos}^2 2\theta$$

and adding the two equations together,

$$\left[\sigma_n - \frac{1}{2}(\sigma_x + \sigma_y)\right]^2 + \tau^2 = \frac{1}{4}(\sigma_x - \sigma_y)^2 \cdot \tau_{xy}^{\ 2} \tag{9.8}$$

Note the right-hand side of Equation (9.8) is a constant quantity for a given system of applied stresses. On inspection the equation can be seen to be of the type

$$(x - a)^2 + y^2 = R^2$$

and is thus the equation of a radius of a circle.

This circle is the graphical representation of Equations (9.5) and (9.7) and thus defines all possible stress conditions.

9.5 CONSTRUCTION OF THE MOHR CIRCLE

The starting point is any two-dimensional stress system, as shown in Figure 9.3(a). In some cases some stresses may be zero. The Mohr circle is fundamentally a graph, and as such it requires horizontal and vertical axes.

The horizontal axis is used to represent direct or axial stresses, and the vertical axis is reserved for shear stresses. If a circle is to be produced, then both axis scales must be identical; otherwise, an ellipse will be generated.

The stresses σ_x and σ_y are marked off along the horizontal axis, and from these points ordinates are projected to equal τ_{xy} and its complementary τ_{yx}. This is shown in Figure 9.6, where these coordinates define the points A and B.

A consistent sign convention needs to be adopted for the stresses, and here it is assumed that σ_x and σ_y will be positive and drawn to the right-hand side of the origin when they represent tensile stresses.

Shear stress systems give rise to a surprising number of sign convention possibilities, and in this instance shear stresses producing a clockwise couple on an element will be taken as positive (see Figure 9.5).

Returning to the construction again, the points A and B just obtained are joined up. The joining line cuts through the base line at M. Now draw a circle, at center M and radius AM.

Examination of Figure 9.6 shows that

$$OM = \frac{\sigma_x + \sigma_y}{2}$$

and that

$$ME = \frac{\sigma_x - \sigma_y}{2}$$

The radius is

$$MA = \sqrt{\left(\frac{\sigma_x - \sigma_y}{2}\right)^2 + \tau_{xy}^{\ 2}}$$

Thus the diagram is analogous to the various related formulas derived earlier.

It is important to note (and remember) that the angle CMB = $2\theta_p$ on the circle diagram corresponds to the angle θ_p on the original stresses element.

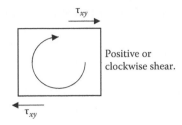

FIGURE 9.5   Sign conventions.

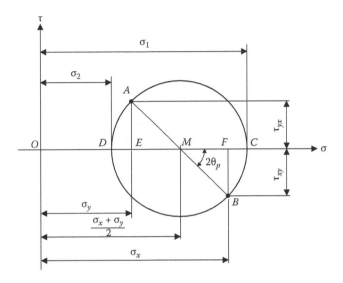

FIGURE 9.6   Mohr circle diagram.

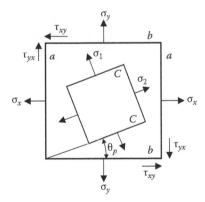

FIGURE 9.7   Stresses acting on the element.

9.5.1   Conclusions and Deductions

The construction of the circle can quickly become an automatic activity. There is no need to consciously use the algebraic equations because the circle provides an alternative.

The two principal stresses σ_1 and σ_2 can be measured immediately. Figure 9.7 shows that there is no shear stress associated with the principal stress. The principal stresses are also clearly seen to be the maximum and minimum values of the normal stress. The half height of the circle gives the maximum shear stress (accompanied, in general, by an axial stress).

Note also that the principal stresses, on the circle, are 180° apart, and also that the maximum shear stress is 90° away from the principal stresses. This signifies that the corresponding angles in the element are 90° and 45°, respectively.

Point c on the diagram represents the maximum principal stress σ_1 and is separated from B by a counterclockwise rotation of $2\theta_p$. On the element the principal plane upon which σ_1 acts is therefore at an angle θ_p counterclockwise from the planes b (Figure 9.7). The plane for the minimum principal stress σ_2 is obtained by turning through a further 90°, since D on the circle is 180° from C.

9.6   RELATIONSHIP BETWEEN DIRECT AND SHEAR STRESS

At this point in the discussion it may be of interest to the reader to show the graphical relationship between the direct stress σ and shear stress τ.

Consider a bar of cross-sectional area A subjected to an axial tensile load P, and it is required to investigate the nature of the stresses on any plane $x{:}x$ making an angle θ with the normal cross section (Figure 9.8).

Let σ_θ and τ_θ be the normal, i.e., direct and tangential (shear), components of the resultant stress on the interface, the directions being assumed to be as shown.

Stress on the normal cross section: $$\sigma = P/A$$

Area of section x:x: $$= A \operatorname{Sec}\theta = A/\operatorname{Cos}\theta$$

Resolving forces perpendicular to x:x: $$\sigma_\theta \times A/\operatorname{Cos}\theta = P \operatorname{Cos}\theta$$

Therefore, $$\sigma_\theta = (P/A) \operatorname{Cos}^2\theta$$

Hence, $$\sigma_\theta = \sigma\operatorname{Cos}^2\theta$$

Since $\operatorname{Cos}^2\theta$ is positive for all values of θ, σ_θ can never be negative, i.e., compressive.

The maximum value of σ_θ occurs when $\operatorname{Cos}^2\theta$ is a maximum, i.e., when $\theta = 0°$. Then $\sigma_\theta = \sigma$. This is the stress on y:y.

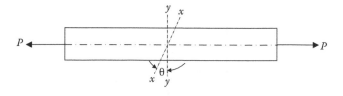

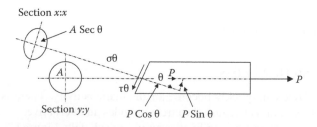

FIGURE 9.8   Section through a prismatic bar.

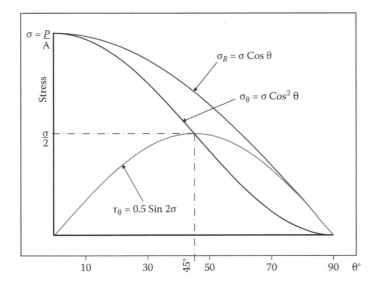

FIGURE 9.9 Distribution of stress from Figure 9.8.

Resolving forces parallel to x:x, $\quad\quad \tau_\theta \times A/\cos\theta = P \sin\theta$

Therefore, $\quad\quad\quad\quad\quad\quad\quad\quad \tau_\theta = P/A \sin\theta \cos\theta$

Hence, $\quad\quad\quad\quad\quad\quad\quad\quad\quad \tau_\theta = 1/2 \, \sigma \sin2\theta$

For values of θ greater than 90° the above expression becomes negative and the direction of τ_θ is the reverse of that assumed.

The maximum value of τ_θ occurs when $\sin2\theta$ is a maximum, i.e., when $\sin2\theta = 1$. Then $2\theta = 90°$, i.e., $\theta = 45°$. Then $\tau_\theta = \sigma/2$.

9.7 THE POLE OF THE MOHR CIRCLE

The method just outlined would be sufficient to answer any two-dimensional stress problem. However, with a very simple addition the circle can be made to indicate directly the axes or planes of the stresses. The circle diagram of Figure 9.6 can be extended by drawing through point B a line parallel to the τ axis and cutting the circle at P_1, which is termed the pole. This point P_1 is the pole for the axes of direct stresses. If a line is drawn from any point on the circle to pass through the pole, then the direction of this line gives the axis for the stress component represented by the point on the circle from which the line originated. This is shown in Figure 9.10.

If instead of drawing lines through A and B parallel to the stress axes, you draw lines parallel to the planes upon which the stresses act, then the pole point P_2 will be formed from which planes of any stress condition can be found. This is shown in Figure 9.11. The two pole points are diametrally opposite of each other, that is, 180° apart on the circle and 90° on the element, as one would expect for planes and axes. These simple ideas will now be used for a number of different examples.

9.7.1 A Few Special Cases

Before moving on to typical practical examples it is useful to examine some simple stress situations, both for the practice of visualizing the conditions as represented by the circle and for the useful place certain of these cases possess as working concepts in stress analysis.

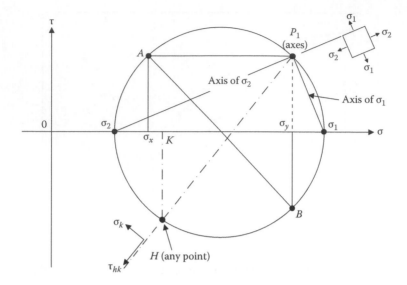

FIGURE 9.10   The pole of the Mohr circle (a).

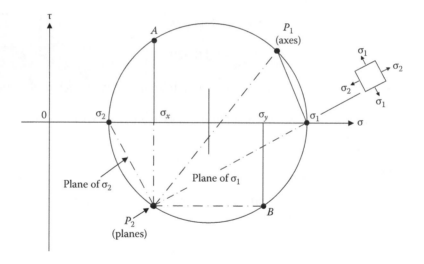

FIGURE 9.11   The pole of the Mohr circle (b).

The following thumbnail sketches in Figure 9.12 will be considered in turn:

(a) and (b): The single stress shown is, of course, a principal stress of zero. Note that the maximum shear stress shown by the circle is $\tau_{max} = 1/2\sigma_y$.

(c): In the example of pure torque, say a shaft, the counterclockwise, and hence negative, shearing stresses shown by the vertical arrows are balanced by the clockwise, therefore positive, complementary shear stresses. These are marked off the vertical axis of the diagram and a circle, center O is drawn to touch them. The circle demonstrates that this shear system gives rise to tensile and compressive stresses equal in magnitude and, as shown by the pole construction, at 45° to the shear planes.

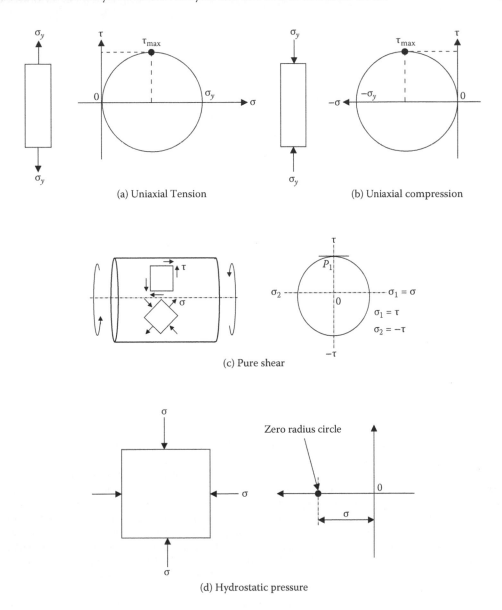

FIGURE 9.12 Types of loading that is treated using the Mohr circle.

(d): This example illustrates the so-called two-dimensional hydrostatic stress system where equal compressive stresses act in the x and y directions. The circle now shrinks to a point showing there is no shear. The case of equal tensile stresses is similar and has important consequences in certain theories of elastic failure.

9.8 EXAMPLES

In this section the ideas so far developed will be used in examples of a more realistic nature. It is hoped that the ease of the method will become apparent and that the reader will be able to make similar analysis for his or her own problems. The first two examples are not taken directly from a practical problem, but indicate the form into which a practical problem must be put before it can be analyzed by the circle, or indeed, by any other method.

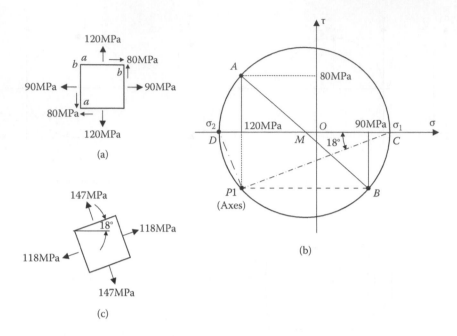

FIGURE 9.13 Numerical example for treatment by Mohr circle.

Example 9.1

Given a complex stress situation, find the principal stresses and their inclination.

The complex stress situation is shown in Figure 9.13(a). Choose a suitable scale for the stress and draw horizontal and vertical axes. Mark off the 90 MPa in the positive direction, and from this position draw in vertically the shear stress of 80 MPa. Note that the shears on the vertical plane are counterclockwise and therefore negative, so that the resulting point B lies below the σ axis.

Similar treatment of the stresses on the horizontal planes gives the point A. AB is a diameter of the basic Mohr's circle, which may therefore be constructed with center M and radius MA. The values of the maximum and minimum principal stresses can be scaled to give +118 MPa and −147 MPa.

A line is drawn from point A parallel to the axis of the 120 MPa stress cutting the circle, and as a check, another line from B parallel to the axis of the 90 MPa stress; then these will meet on the circle at P1, which is the pole point for the axes.

If the principal stress points C and D are joined to P_1, CP_1, and DP_1 give the directions of the axes of σ_1 and σ_2, respectively. The angle DCP_1 gives the inclination of the axis of maximum principal stress to that of the 90 MPa stress in the original system and is found by measurement to be approximately 18°.

If it had been necessary, a pole point P_2 for planes could have been drawn just as easily. It would appear at the opposite end of the diameter to P_1.

Example 9.2

Find the stress components on any plane when the principal planes and stresses are given.

Figure 9.14(a) shows two principal stresses, and it is required to investigate the value of the stresses on a given line or plane, 40° from the vertical plane.

On the σ axis measure off the two principal stresses. In this example note that these provide the circle extremities because the shear stress component is zero. Find the midpoint of the distance AB and draw in a circle. The construction for the pole of the planes P_2, as indicated in Figure 9.10, shows that in this case P_2 coincides with B. Draw a line through P_2 parallel to the required plane.

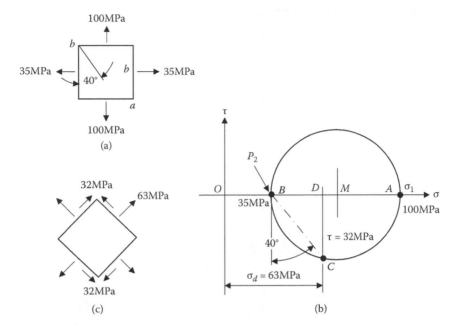

FIGURE 9.14 Example.

This cuts the circle at C, which has coordinates OD and DC. OD is the normal stress on the plane and is found to be 63 MPa, positive and hence tensile. DC is the shear stress and is −32 MPa; consequently, it acts counterclockwise, as shown in Figure 9.14(c).

Example 9.3

The loading on a shaft is such that it causes simultaneous bending and twisting. It is required to calculate the stresses at the surface of the shaft due to these moments acting separately and then using the Mohr's circle to find the principal stresses and planes for the combined loading.

Finally, to calculate the factor of safety based on a given yield stress, use the Von Mises shear stress-strain energy criterion.

Data
Bending moment M	= 1,600 Nm
Twisting moment T	= 850 Nm
Yield in pure tension	= 300 Mpa
Diameter d	= 65 mm
Factor of safety (FoS)	To be determined
Material	Mild steel

In the following calculations the simple theory of bending and torsion will be assumed. The appendix briefly describes the main theories of elastic failure.

Calculation of the Bending Stress

$$I_{xx} = \frac{\pi \cdot d^4}{64}$$

$$I_{xx} = 876240.51\,mm^4$$

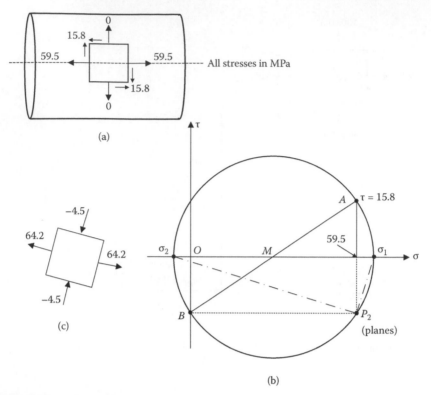

FIGURE 9.15 Example–loading on a shaft.

Now applying the beam theory:

$$\sigma_x = \frac{M \cdot d}{I_{xx} \cdot 2}$$

$$\sigma_x = 59.34 \, \text{MPa}$$

Calculation for the shear stress due to torsion:

$$J = \frac{\pi \cdot d^4}{32}$$

$$J = 1752481.01 \, \text{mm}^4$$

Now applying the torsion theory:

$$T_{xy} = \frac{\tau \cdot d}{J \cdot 2}$$

$$T_{xy} = 15.76 \, \text{MPa}$$

The system of maximum stresses at the shaft surface is shown in Figure 9.14(a), in which it should be noted that the circumferential direct stress is zero and that the bending stress is taken as tensile. An identical system will exist at the opposite end of the shaft, except the bending stress will be compressive.

The above values are now used to construct the circle in Figure 9.14(b); from the points A and B lines are drawn parallel to the appropriate planes to obtain a pole point P_2. The principal stresses are measured and found to be 64.2 MPa tensile and 4.5 MPa compressive. The principal stress points on the circle are joined up to the pole P_2 and the principal plane diagram is drawn in Figure 9.14(c). Carefully note that this is a pole for planes, and the axial stresses are at right angles to the plane. It would help to fix ideas if the reader would sketch in a pole P_1 for the axes of the direct stresses and show that it produces an identical result.

To Find the Factor of Safety (FoS)

The Von Mises theory for a two-dimensional stress system is given by

$$\sigma_y^2 = \sigma_1^2 + \sigma_2^2 - \sigma_1 \cdot \sigma_2$$

where σ_1 and σ_2 are the principal stresses at which failure will occur if the yield stress in pure tension is σ_y.

In this case there is a Factor of Safety (FoS) = σ_y / σ_w where σ_w is the working stress given by:

$$\sigma_w^2 = \sigma_1^2 + \sigma_2^2 - \sigma_1 \cdot \sigma_2$$

$$\sigma_w = 66.564 \text{ MPa}$$

$$\text{FoS} = \frac{\sigma_y}{\sigma_w}$$

$$\text{FoS} = 4.507$$

If the circle construction were applied at the opposite end of the shaft, where the bending stress is compressive, the principal stresses would be numerically identical but opposite in sign, so that σ_w and consequently FoS would have the same value as above.

Example 9.4

Figure 9.16(a) shows a cross section of a universal beam 416 mm × 154 mm × 74 kg/m³. The other dimensions relative to the problem are shown on the diagram.

A vertical shear force of 140 kN is applied to the section together with a bending moment of 100 kNm, which acts simultaneously with the shear force. The second moment of area is 270 × 10⁶ mm⁴.

In the first instance it is required to estimate the maximum principal stress, which would occur just above the flange in the web.

Second, if the maximum principal stress is limited to 100 MPa, find how much the shear stress and force could be increased, if the bending moment is maintained constant.

Figure 9.16(b) shows the element under consideration. There will be a tensile stress due to bending and shearing stresses due to the vertical shear force. These are calculated in the following section.

The stress due to bending can be calculated from the beam theory:

$$M = 100 \times 10^3 \text{ Nm}$$

$$y = 190 \text{ mm}$$

$$I = 270 \times 10^6 \text{ mm}^4$$

$$\sigma_b = \frac{My}{I}$$

$$\sigma_b = 70.37 \frac{N}{mm^2}$$

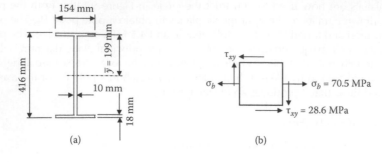

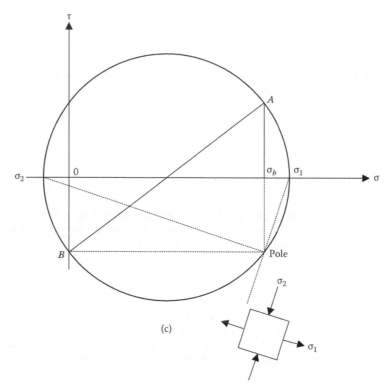

FIGURE 9.16   Example–universal beam part 1.

For the shear stress at the required point the following relationship is used:

$$Q = 140 \times 10^3 \text{ N (vertical shear force)}$$
$$I = 270 \times 10^6 \text{ mm}^4 - \text{second moment of area}$$
$$A = 2772 \text{ mm}^2 \text{ (area of section above the line upon which shear stress is required)}$$
$$y = 199 \text{ mm (vertical distance of centroid of this area from neutral axis line)}$$
$$b_o = 10 \text{ mm (width of section at the point where shear stress is required)}$$
$$\tau_{xy} = \frac{Q}{I \cdot b_o}(Ay)$$
$$\tau_{xy} = \frac{140 \times 10^3 \text{ N} \times 154 \text{mm} \times 18 \text{m} \times 199 \text{m}}{270 \times 10^6 \text{ mm}^4 \times 10 \text{ mm}}$$
$$\tau_{xy} = 28.6 \ \frac{\text{N}}{\text{mm}^2}$$

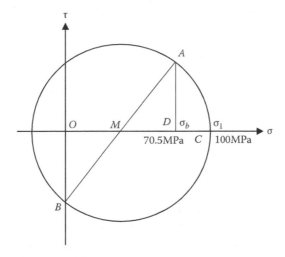

FIGURE 9.17 Example–universal beam part 2.

For the stress system shown the Mohr circle has been constructed in accordance with Figure 9.16(c). Point A represents the stress condition on the vertical face and point B that on the horizontal face of the element. For completeness the pole point for the planes has been obtained and the orientation of the principal stress system shown. The value of the maximum principal stress is 80.6 N/mm².

The Mohr circle for the second part of the problem is shown in Figure 9.17. The bending moment, and hence the stress due to bending, does not change. The maximum principal stress is given so the extremity of the circle can be placed on the horizontal axis at C. The center of the circle must lie halfway between the origin O and the point D representing σ_b. The point M is the center of the circle.

Thus the circle of radius MC can be drawn. Points A and B can now be obtained and the allowable shear stress read off the diagram.

New shear stress = 53.9 MPa

Hence the new value of shear force can be obtained, pro rata, from the original value.

$$Q = \frac{140 \times 53.9}{28.6}$$

$$= 264 \text{ kN}$$

9.9 THE ANALYSIS OF STRAIN

Previous sections have shown the basis for the use of the Mohr circle when solving complex stress problems. However, in some areas of analysis the starting point may be data recorded as strains generally obtained from electrical resistance strain gauges of one kind or another.

Ultimately this information has to be processed and converted into information about stresses. Therefore, the next section will derive a set of equations relating to strains.

9.9.1 Sign Conventions for Strains

For linear strain, extensions are considered positive. For the case of shear strain it is desirable to choose a system that is simple to use and is fully consistent with the stress conventions already adopted.

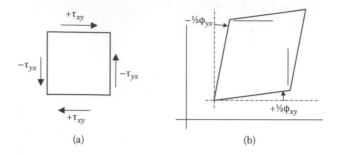

(a)   (b)

FIGURE 9.18 Sign conventions for strain.

Figure 9.18(a) and (b) shows a pure shear stress system with the signs added in accordance with the conventions to be used here. The deformation of the element is shown and the signs are chosen for the semishear strains to correspond to those used for the stresses.

From the early sections the reader will recall that the first letter in the double suffix notation refers to the plane and the second letter gives the direction of the normal to the plane on which the stress acts.

Note that now τ_{xy}, which is positive, accompanies $\tfrac{1}{2}\varphi_{xy}$, which is a counterclockwise rotation and is also taken as positive. This has the added advantage of agreeing with the standard mathematical usage. The other rotation of $\tfrac{1}{2}\varphi_{yx}$ is negative and corresponds with the stress conditions.

It can be shown that the equations for transforming the strains are as follows:

$$\varepsilon_x^1 = \frac{\varepsilon_x + \varepsilon_y}{2} + \frac{\varepsilon_x - \varepsilon_y}{2}\,Cos2\theta + \frac{\gamma_{xy}}{2}\,Sin2\theta \tag{9.9}$$

$$\varepsilon_y^1 = \frac{\varepsilon_x + \varepsilon_y}{2} + \frac{\varepsilon_x - \varepsilon_y}{2}\,Cos2\theta - \frac{\gamma_{xy}}{2}\,Sin2\theta \tag{9.10}$$

$$\gamma_{xy}^1 = \frac{-\left(\varepsilon_x - \varepsilon_y\right)}{2}\,Sin2\theta + \frac{\gamma_{xy}}{2}\,Cos2\theta \tag{9.11}$$

9.10 COMPARISON OF STRESS AND STRAIN EQUATIONS

$$\sigma_n = \frac{1}{2}\left(\sigma_x + \sigma_y\right) + \frac{1}{2}\left(\sigma_x - \sigma_y\right)Cos2\Theta - t_{xy}Sin\Theta \quad \text{From equation (9.1)}$$

$$\varepsilon_a = \frac{1}{2}\left(\varepsilon_x + \varepsilon_y\right) + \frac{1}{2}\left(\varepsilon_x - \varepsilon_y\right)Cos2\Theta - 2\varphi_{xy}Sin2\Theta \tag{9.12}$$

and

$$\tau = \frac{1}{2}\left(\sigma_x - \sigma_y\right)Sin2\Theta + t_{xy}Cos2\Theta \quad \text{From equation (9.2)}$$

$$\frac{1}{2\varphi_{ab}} = \frac{1}{2}\left(\varepsilon_x - \varepsilon_y\right)Sin2\Theta - \frac{1}{2}\varphi_{xy}Cos2\Theta \tag{9.13}$$

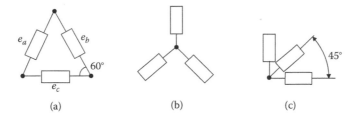

FIGURE 9.19   The strain rosette.

The stress and strain equations are seen to be completely analogous except that the strain quantity corresponding to τ_{xy} is $1/2\varphi_{xy}$ and not φ_{xy}.

Provided the vertical axis is represented by $1/2\varphi$ the circle can be represented by a circle diagram.

The construction of the strain circle will be identical to the stress circle if the linear and shear strains are part of the data.

9.10.1   THE STRAIN ROSETTE

One of the most common methods of measuring strain on a component is to use an electrical resistance strain gauge. They can only measure tensile or compressive strains but are not able to measure shearing strains. When the direction of the principal strains is not known, it is usual practice to fit a strain rosette consisting of three gauges, which are displaced with an angular relationship to each other. Figure 9.19 shows some arrangements of strain gauge rosettes.

9.10.2   CONSTRUCTION

1. Select suitable scales for the linear strain axis (abscissa) and half the shear strain (ordinate).
2. Considering Figure 9.19(c). The points representing the strains ε_1 and ε_3 must lie on the circumference of the circle and be 180° apart (since grids 1 and 3 are 90° apart). To establish the center of the circle, algebraically add ε_1 and ε_3, divide by 2, and plot the result, observing the sign on the abscissa.
3. To establish the radius, the magnitude of $\varphi_{1,3}$ is determined by $\varepsilon_2 - (\varepsilon_1 + \varepsilon_3)/2$; the value is laid off from ε_1 (observing the sign) parallel to the ordinate. The circle is then drawn.
4. Values for ε_{max} and ε_{min} occur where the circle crosses the ε axis (since the shearing strains are zero at these points). γ_{max} is the diameter of the circle on the $\gamma/2$ scale (i.e., twice the radius). The angle of the maximum principal strain is that subtended by the abscissa and the radius (where $\gamma_{1,3}$ meets the circle), divided by 2, the sign according to the code illustrated.

The Mohr's strain circle can also be constructed for strains measured at angles other than 45°. The method is due to F. A. McClintock.

Let the measured strains be orientated as shown below.

The strains (ε) can have any value including the angles (θ).

To draw the Mohr's strain circle:

1. The rosette axes are arranged by extending, if necessary, so that they are:
   a.   Arranged in sequence, i.e., in order of ascending or descending order of strain magnitudes
   b.   The included angles between the axes of the maximum and minimum strains are less than 180°

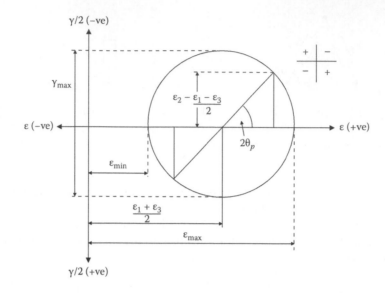

FIGURE 9.20   Mohr strain circle.

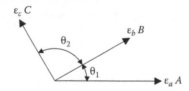

FIGURE 9.21   Rosette strains.

If α = angle between the max and strain axes, and β = angle between the intermediate and min strain axes, rearrange the layouts such that the intermediate strain is in the vertical position.

In case 3 the axis of max strain falls to the left of the intermediate, and hence further extensions will be necessary.

2. Plot the strains as shown without inserting an x axis.
3. Construct a circle through points DEF. This is Mohr's strain circle. Now draw X axis.

For the case where $\varepsilon_a > \varepsilon_b > \varepsilon_c$, ε_1 and ε_2, and the principal strains, the points E, A, and F represent ε_a, ε_b, and ε_c, respectively. They must be chosen to satisfy the following requirements:

1. The magnitudes of εa, εb, and εc.
2. The rotational sequence must correspond to the rosette layout.

Principal stresses are given by

$$\sigma_1 = \frac{E\left(\varepsilon_1 + v\varepsilon_2\right)}{1 - v^2}, \qquad \sigma_2 + \frac{E\left(\varepsilon_2 + v\varepsilon_1\right)}{1 - v^2}$$

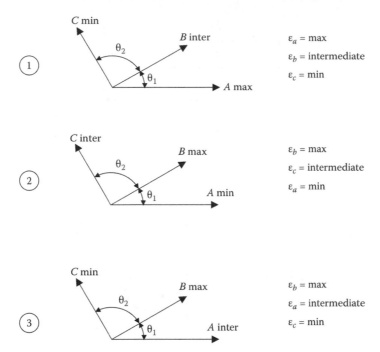

FIGURE 9.22 Rosette axes.

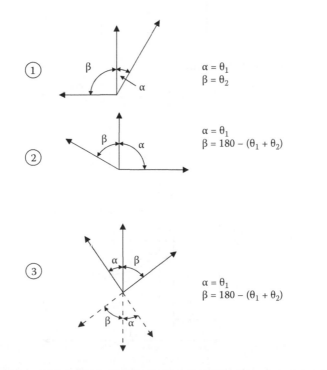

FIGURE 9.23 Rearrangement of rosette axes

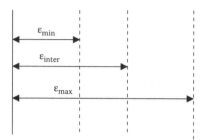

FIGURE 9.24   Strain plot.

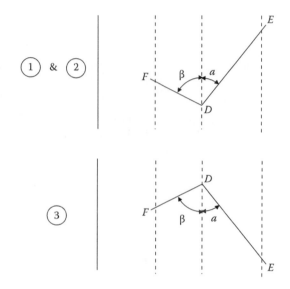

FIGURE 9.25   Construction of Mohr strain circle.

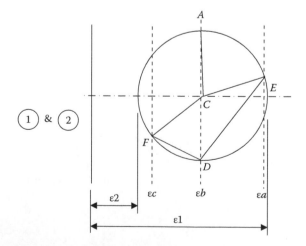

FIGURE 9.26   Construction of Mohr strain circle (continued).

9.10.3 CONCLUSION

This concludes the introduction to the construction of the Mohr's circle, and the author hopes that it has given the student an understanding of the construction and use of the circle in solving complex stress problems, together with the use of the circle in transposing strains derived from electrical resistance strain gauges to principal stresses.

There are other areas where the circle is useful: notably in certain flat plate problems and soil mechanics. These topics are rather specialized and have not been treated in this text.

However, a familiarity with the uses advocated here should make it a comparitively easy task to extend its use in these and other topics.

9.11 THEORIES OF ELASTIC FAILURE

9.11.1 STEADY LOAD FAILURE THEORIES

Elastic failure is assumed to occur when the internal stresses in the test piece reach the elastic limit stress for the material, which will be denoted by σ_0 when subjected to a simple tension test. When a component is subjected to a complex stress system, elastic failure may not necessarily occur when the greatest principal stress reaches the elastic limit stress σ_0. The other (lesser) principal stresses in the orthogonal directions may affect the limiting value of the greatest principal stress at failure, which may then be greater or lesser than σ_0.

The effect of the lesser principal stresses depends upon a number of factors and applications, i.e., such as if the applied forces are in the same or opposite directions to the greatest principal stress and whether the material is either ductile or brittle.

Various theories have been proposed on the elastic failure of a material under complex stress, and these are usually associated with the name of the originator. Substantial research has been undertaken on these theories, and the most important applications are given after each theory.

The system of stresses applied to a component can be resolved into three principal stresses, as shown in Figure 9.27 and in the following. It is assumed that $\sigma_x > \sigma_y > \sigma_z$ and that they are all the same sign.

9.11.2 MAXIMUM PRINCIPAL STRESS (RANKIN'S) THEORY

Failure will be considered to have occurred when one of the principal stresses reaches the elastic limit stress in a simple tensile stress test, irrespective of the other principal stresses (see Figure 9.28).

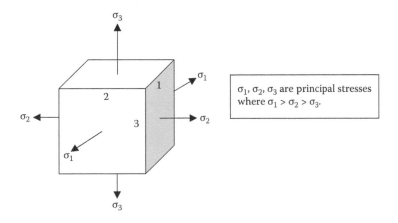

FIGURE 9.27 Principle orthogonal principal stresses.

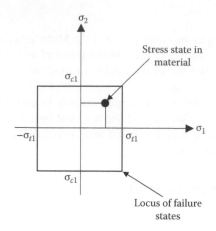

FIGURE 9.28 Maximum principal stress (Rankin's) theory.

Failure occurs when

$$\sigma_1 = \sigma_{t1} \qquad \sigma_{t1} = \text{yield strength in tension}$$

$$\sigma_3 = \sigma_{c1} \qquad \sigma_{c1} = \text{yield strength in compression}$$

In a biaxial stress condition the theory implies that if the stress state falls within the boundary depicted in Figure 9.28, the material will not fail.

The theory has been found to be approximately true for brittle materials but not for ductile materials. Rupture will usually take place on a plane inclined to the plane of greatest direct stress, indicating that failure is due to shear.

9.11.3 MAXIMUM PRINCIPAL STRAIN (ST. VENANT'S) THEORY

This theory predicts that failure will occur when the greatest of the three principal strains becomes equal to the strain corresponding to the yield strength (see Figure 9.29), i.e.,

$$E\varepsilon_1 = \sigma_1 - v(\sigma_2 + \sigma_3) = \pm\sigma_y$$

$$E\varepsilon_2 = \sigma_2 - v(\sigma_1 + \sigma_3) = \pm\sigma_y$$

$$E\varepsilon_3 = \sigma_3 - v(\sigma_1 + \sigma_2) = \pm\sigma_y$$

E = Modulus of elasticity.
v = Poisson's ratio

For a biaxial condition: (plane stress)
ie.,

$$\sigma_3 = 0$$

$$\sigma_1 v \sigma_2 = \pm\sigma_y$$

$$\sigma_2 v \sigma_1 = \pm\sigma_y$$

as long as the stress state falls within the polygon, the material will not yield.

This theory is not substantiated by experiment and finds little general support.

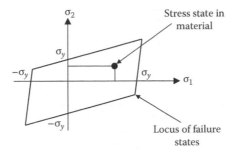

FIGURE 9.29   Maximum principal strain (St. Venant's) theory.

9.11.4   Maximum Shear Stress (Guest's or Tresca's) Theory

Yielding begins whenever the maximum shear stress in a part becomes equal to the maximum shear stress in a tension test specimen that begins to yield.

Yielding will occur if any of the following criteria are met (Figure 9.30):

$$\pm\sigma_y = \sigma_1 - \sigma_2$$

$$\pm\sigma_y = \sigma_2 - \sigma_3$$

$$\pm\sigma_y = \sigma_1 - \sigma_3$$

For the biaxial case (plane strain) $\sigma_3 = 0$.

$$\pm\sigma_y = \sigma_1 - \sigma_2$$

$$\pm\sigma_y = \sigma_2$$

$$\pm\sigma_y = \sigma_1$$

Note: In quadrants I and III the Maximum Principal Stress Theory and the Maximum Shear Stress Theory are the same for the biaxial case.

This theory gives good correlation with experimental results obtained from ductile material.

9.11.5   Distortion Energy Theory

9.11.5.1   Strain Energy (Haigh's) Theory

Yielding will occur when the distortion energy per unit volume equals the distortion energy per unit volume in a uniaxial tension specimen stressed to its yield strength (see Figure 9.31).

The strain energy per unit volume is given by the equation

$$U = \frac{1}{2}\sigma_1 \cdot \varepsilon_1 + \frac{1}{2}\sigma_2 \cdot \varepsilon_2 + \sigma_3 \cdot \varepsilon_3$$

Units :

$$[U] = [N/mm2] \ *[mm/mm] = [N.mm/mm^3]$$

An expression for the strain energy per unit volume in terms of stress only can be obtained by making use of the stress-strain relationship.

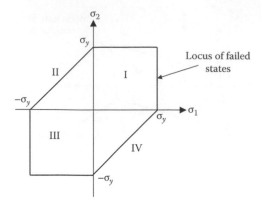

FIGURE 9.30 Maximum shear stress (Guest or Tresca's) theory. Note: In quadrants I and III the maximum principal stress theory and the maximum shear stress theory are the same for the biaxial case.

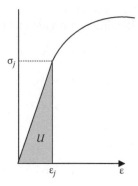

FIGURE 9.31 Strain energy (Haigh's) theory.

In algebraic format:

$$\varepsilon_1 = \frac{1}{E}\left(\sigma_1 - \nu\sigma_2 - \nu\sigma_3\right)$$

$$\varepsilon_2 = \frac{1}{E}\left(\sigma_2 - \nu\sigma_1 - \nu\sigma_3\right)$$

$$\varepsilon_3 = \frac{1}{E}\left(\sigma_3 - \nu\sigma_1\nu\sigma_2\right)$$

$$U = \frac{1}{2}\cdot\sigma_1\cdot\varepsilon_1 + \frac{1}{2}\cdot\sigma_2\cdot\varepsilon_2 + \frac{1}{2}\cdot\sigma_3\cdot\varepsilon_3$$

$$= \frac{1}{2}\sigma_1\left[\frac{1}{E}\cdot\left(\sigma_1 - \nu\sigma_2 - \nu\sigma_3\right)\right] + \frac{1}{2}\sigma_2\left[\frac{1}{E}\cdot\left(\sigma_2 - \nu\sigma_1 - \nu\sigma_3\right)\right] + \frac{1}{2}\sigma_3\left[\frac{1}{E}\cdot\left(\sigma_3 - \nu\sigma_1 - \nu\sigma_2\right)\right]$$

$$U = \frac{1}{2E}\left(\sigma1^2 + \sigma2^2 + \sigma3^2 - 2\nu\left(\sigma_1\cdot\sigma_2 + \sigma_2\cdot\sigma_3 + \sigma_3\cdot\sigma_1\right)\right)$$

In matrix format:

$$\begin{Bmatrix} \varepsilon_1 \\ \varepsilon_2 \\ \varepsilon_3 \end{Bmatrix} = \begin{pmatrix} 1 & -v & -v \\ -v & 1 & -v \\ -v & -v & 1 \end{pmatrix} \begin{Bmatrix} \sigma_1 \\ \sigma_2 \\ \sigma_3 \end{Bmatrix}$$

This theory is well supported by experimental results on ductile materials, particularly with thick cylinders. It breaks down in the case of hydrostatic pressure ($\sigma_1 = \sigma_2 = \sigma_3$):

It predicts failure when

$$\sigma = \frac{\sigma_o}{\sqrt{3 \cdot (1 - 2v)}}$$

when in fact no failure would occur.

9.11.5.2 Shear Strain Energy (Von Mises's) Theory

This theory is known by a number of names.

- Maximum shear strain energy per unit volume theory
- Distortion energy theory
- Von Mise's-Hencky theory
- Maximum octahedral shear stress theory

This theory is commonly known as Von Mise's theory.

Failure is predicted to occur when the shear strain energy stored per unit volume in a strained material reaches the shear strain energy per unit volume at the elastic limit in a simple tension test. This is similar to the previous theory, but it assumes that the volumetric strain energy plays no part in producing elastic failure.

Thus

$$\frac{1}{6G} \left[\sigma_x^2 + \sigma_y^2 + \sigma_z^2 - (\sigma_x \cdot \sigma_y + \sigma_y \cdot \sigma_z + \sigma_z \cdot \sigma_x) \right] = \frac{\sigma_o}{6G}$$

or

$$\sigma_x^2 + \sigma_y^2 + \sigma_z^2 - (\sigma_x \cdot \sigma_y + \sigma_y \cdot \sigma_z + \sigma_z \cdot \sigma_x) = \sigma_o^2$$

This theory has received considerable verification in practice and is widely regarded as the most reliable basis for design.

9.11.6 CONCLUSIONS

It has been found that both the distortion energy theory and the maximum shear stress theory provide reasonable estimates for the onset of yielding in the case of static loading for ductile, homogeneous, isotropic materials whose compression and tensile strengths are the same.

Both distortion energy theory and the maximum shear stress theory predict the onset of yielding and are independent of hydrostatic stress. This agrees reasonably well with experimental data for moderate hydrostatic pressures.

Both the distortion energy theory and the maximum shear stress theory underpredict the strength of brittle materials loaded in compression. Brittle materials often have much higher compressive strengths than tensile strengths.

The distortion energy theory is slightly more accurate than the maximum shear stress theory. The distortion energy theory is the yield criteria most often used in the study of plasticity. Its continuous nature makes it more mathematically amenable.

9.12 INTERACTION CURVES, STRESS RATIO'S MARGINS OF SAFETY, AND FACTORS OF SAFETY

9.12.1 INTERACTION: STRESS RATIO

Yielding or failure of a structural member subjected to combined stresses may be predicted without using principal stresses by using what is known as the interactive method.

The method represents applied and allowable stress conditions on a structural member by stress ratios. Stress ratios are nondimensional coefficients R given by

$$R = \frac{applied\ stress}{allowable\ stress}$$

Load ratios may also be used instead of the stress ratios if this is more convenient.

The method involves determining the allowable stress or loading for each separate failure mode, such as tension, compression, bending, buckling, and shear. Stress ratio R for each separate failure condition is calculated and combined using interactive equations, if these loads act simultaneously on the member.

The equation generally takes the following form:

$$R_1^x R_2^y + R_3^z \ldots = 1.0$$

where x, y, z are exponents defining interactive relationships.

Failure is when the sum of the stress ratios is greater than 1.0.

9.12.2 INTERACTIVE CURVE

Interactive curves are used, for convenience, to show the relationship given by the interaction equations (see Figure 9.32); e.g.,

Point A is located with the coordinates R_1 and R_2.

The margin of safety assuming each load increases proportionately until failure occurs at point B.

$$MS = \frac{OB}{OA} - 1$$

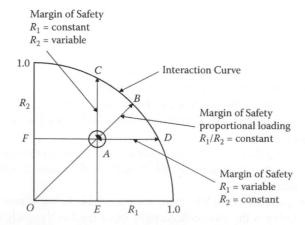

FIGURE 9.32 Interaction curve.

The margin of safety (MS) assuming R_1 remains constant with R_2 increasing until failure occurs at point C.

$$MS = \frac{EC}{EA} - 1$$

The margin of safety assuming R_2 remains constant with R_1 increasing until failure occurs at point D.

$$MS = \frac{FD}{FA} - 1$$

9.12.3   INTERACTION, STRESS RATIOS, YIELD CONDITIONS

To investigate yielding on a compact structure (crippling and buckling not pertinent) under the action of two-dimensional combined stresses, the following method may be used, as an alternative to calculating principal stresses.

A maximum equivalent Stress Ratio $\bar{R}$ is computed and the MS (yield) obtained there from:

$$MS(\text{yield}) = \frac{1}{\bar{R}} - 1$$

R is given by the following formula (which is based on the maximum shear stress energy per unit volume theory):

$$R = \left[R_{n1}^2 + R_{n2} - R_{n1}R_{n2} + R_s^2 \right]^{0.5}$$

where $R_n = R_t + R_b$ or $R_c + R_b$ and subscripts 1 and 2 indicates mutually perpendicular directions.

Stress ratios due to direct stress are positive for tension and negative for compression. The effect of bending stress may be added to the above, ensuring consistency of signs; i.e., bending stresses are positive for tension and negative for compression.

$$R_t = \frac{f_t}{\bar{F}_{ty}}$$

$$R_c = \frac{f_c}{\bar{F}_{cy}}$$

$$R_s = \text{shear stress ratio} = R_s = \frac{f_s}{\bar{F}_{sy}}$$

where $\bar{F}_{sy} =$ may be taken as $0.577 \cdot \bar{F}_{ty}$.

9.12.4 INTERACTION EQUATIONS: YIELD CONDITIONS

TABLE 9.1

Interaction Equations – Yield Conditions

Type of Combined Loading	Tension Stresses +ve Compression Stresses –ve

Uniaxial Tension and Shear

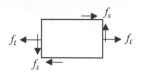

$$R_t = \frac{f_t}{\overline{F}_{ty}} \qquad\qquad R_s = \frac{f_s}{\overline{F}_{sy}}$$

Interaction formula $= R_t^2 + R_s^2 = 1$

$$M.S. = \frac{1}{\left(R_t^2 + R_s^2\right)^{0.5}} - 1$$

Uniaxial Compression and Shear

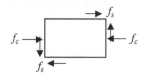

$$R_c = \frac{f_c}{\overline{F}_{cu}} \qquad\qquad R_s = \frac{f_s}{\overline{F}_{sy}}$$

Where $\overline{F}_{cu} = 2\overline{F}_{sy}$

Interaction formula $= R_c^2 + R_s^2 = 1$

$$M.S. = \frac{1}{\left(R_c^2 + R_s^2\right)^{0.5}} - 1$$

Biaxial Tension

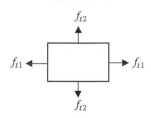

No interaction between axes

$$M.S. = \frac{1}{R} - 1$$

R is greater of R_{t1} and R_{t2}

$$R_{t1} = \frac{f_{t1}}{\overline{F}_{ty1}} \qquad R_{t2} = \frac{f_{t2}}{\overline{F}_{ty2}}$$

Biaxial Compression

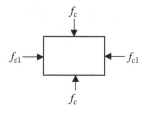

No interaction between axes

$$M.S. = \frac{1}{R} - 1$$

R is greater of R_{c1} and R_{c2}

$$R_{c1} = \frac{f_{c1}}{\overline{F}_{cy1}} \qquad R_{c2} = \frac{f_{c2}}{\overline{F}_{cy2}}$$

Biaxial Normal Stresses and Shear

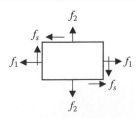

$R_{n1} = R_{t1}$ or R_{c1} $\qquad\qquad R_{n2} = R_{t2}$ or R_{c2}

Interaction Formula

$$= R_{n1}^2 + R_{n2}^2 - R_{n1}^2 R_{n2}^2 + R_s^2 = 1$$

$$M.S. = \frac{1}{\left[R_{n1}^2 + R_{n2}^2 - R_{n1}^2 R_{n2}^2 + R_s^2\right]^{0.5}} - 1$$

9.12.5 INTERACTION EQUATIONS: FAILURE CONDITIONS

9.12.5.1 Compact Structures: No Bending

The following interaction equations for failure conditions are similar to those for the yield conditions, as shown, except for the following:

where:

$$R_t = \frac{f_t}{\overline{F}_{tu}}$$

$$R_c = \frac{f_c}{\overline{F}_{cu}}$$

$$R_s = \frac{f_s}{\overline{F}_{su}}$$

where:

$\overline{F}_{tu}$ = ultimate allowable tensile strength of the material.

$\overline{F}_{cu}$ = ultimate allowable compressive strength of the material.

$\overline{F}_{su}$ = ultimate allowable shear strength of the material.

9.12.6 COMPACT STRUCTURES: BENDING

1. Bending within the elastic limit. The stresses resulting from bending, i.e., f_t and f_c, and the resulting stress ratios should be treated as per the yield conditions.
2. Bending in the plastic range. Pure bending in the plastic range is covered in Chapter 1, where the allowable bending moment $\overline{M}_b$ is evaluated.

$$R_s = \frac{M_b}{\overline{M}_b}$$

where
M_b is the applied bending moment
$\overline{M}_b$ is the allowable bending moment

3. Plastic bending and flexural shear.

$$R_s = \frac{f_s}{f_{su}}$$

where
f_s is the applied bending moment
$\overline{F}_{su}$ is the allowable bending moment

Interaction equation

$$\approx R_b^2 + R_s^2 = 1$$

$$M.S. = \frac{1}{\left[R_b^2 + R_s^2\right]0.5} - 1$$

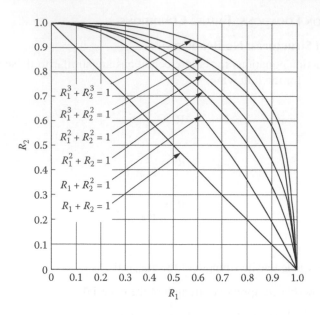

FIGURE 9.33   General interactive relationships.

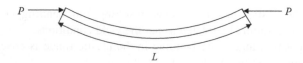

FIGURE 9.34   Direct compression and bending.

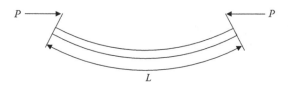

FIGURE 9.35   Offset compression and bending.

9.12.7   GENERAL INTERACTION RELATIONSHIPS (FIGURES 9.33–9.36)

Example 9.5
Consider a structural member subject to a combined axial compression and bending.
   In this case, the axial compression will cause additional deflection, which in turn increases the moment of the bending load. This increase can easily be taken care of by an amplification factor k.

$$k = \frac{1}{1 - \dfrac{P}{P_{cr}}}$$

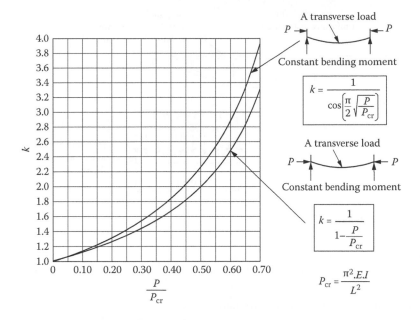

FIGURE 9.36 Amplification factor "k" for bending moment.

where

$$P_{cr} = \frac{\pi^2 E I}{L^2}$$

$$k = \frac{1}{\cos\left(\frac{\pi}{2}\sqrt{\frac{P}{P_{cr}}}\right)}$$

The bending moment applied to the structural member (chosen at the maximum cross section) is then multiplied by the amplification factor k, and this value is then used as the applied moment M in the ratio:

$$R_b = \frac{M}{M_u}$$

The chart in Figure 9.36 is then used to determine the amplification factor k for the bending moment applied to a beam when it is also subject to axial compression.

The resulting combined stress is then found from the following:

$$\sigma = \frac{P}{A} \pm \frac{kM_c}{I}$$

Example 9.6
A loading platform is to be fabricated from a 10.0 mm thick top plate and an under frame manufactured from 25 mm equal angle. The whole structure is in the form of a truss (Figure 9.37).

It is required to determine the compound stress (axial compression and bending) in the top compression panel.

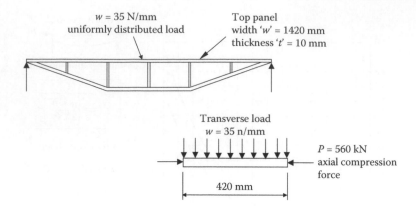

FIGURE 9.37   Truss diagram.

Summary

L = 420.0 mm $I = \dfrac{W.t^2}{12}$

W = 1420.0 mm $I = 118.333 \times 10^3$ mm⁴

t = 10.0 mm

w = 35.0 N / mm   A = W.t

P = 560.0 kN $A = 14.2 \times 10^3$ mm²

E = 206.842 GPa

Critical load $\left(P_{cr}\right)$:

$P_{cr} = \dfrac{p^2.E.I}{L^2}$

$Pcr = 1.369 \times 10^6 N$

Ratio:

$\dfrac{P}{P_{cr}} = 0.409$

Bending moment:

$M = \dfrac{w.L^2}{8}$

$M = 771.75$ kN.mm

Obtaining the amplification factor k for the sinusoidal bending moment from the curve (Figure 9.36):

$$k = 1.692$$

The actual applied moment due to the extra deflection is found to be

$$M_{actual} = k \cdot M$$

$$M_{actual} = 1.692 \times 771.75 \, kN \cdot mm$$

$$= 1.306 \times 10^6 \, N \cdot mm$$

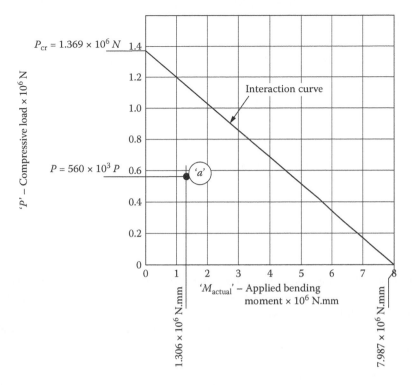

FIGURE 9.38 Interaction curve for truss beam example.

The resulting combined stress above the neutral axis of the top panel is:

$$\sigma_c = \frac{P}{A} + \frac{M_{actual} \cdot t}{I \cdot 2}$$

$$\sigma_c = 94.611 \text{ MPa}$$

$$\sigma_L = \frac{P}{A} - \frac{M_{actual} \cdot t}{I \cdot 2}$$

$$\sigma_L = -15.738 \text{ MPa}$$

9.12.8 DETERMINATION OF SAFETY FACTORS

The ultimate load values for this structural member are in compression alone.

For the values for bending alone, i.e., without the combined compression, the following is to be used.
Assuming material properties, aluminum plate to BS 1470 5154A-H4,

$$\sigma_{0.2} = 225 \text{ MPa}$$

For compression alone:

$$\text{Since } \frac{L}{r} = 150 \qquad (\text{where } r = \text{radius of gyration})$$

$$(\text{assume}) \; P_u = P_{cr}$$

For bending alone:

The plastic or ultimate bending moment is:

$$Mu = \left(W \cdot \sigma_{0.2} \cdot \frac{t}{2} \right) \frac{t}{2}$$

$$Mu = 7.987 \times 10^6 \; N \cdot mm$$

These ultimate values are represented on the interactive curve shown in Figure 9.38.

Plotting the present load values indicates that there is an approximately 2.4:1 factor of safety before the top panel will begin to buckle.

If this had been an extremely short plate (very low L/r ratio), the critical value (P_{cr}) could be much higher than the actual ultimate value (P_u).

10 Introduction to Experimental Stress Analysis

It is important that design calculations are verified, particularly when the design is a complex structure such as an air vehicle wing box or fuselage. A number of factors are used in the analytical analysis, essentially to compensate for lack of specific knowledge generally associated with materials. These factors are referred to as factors of ignorance or "fudge" factors. These factors are used essentially to enable the design to proceed with the analysis being revisited when there is better information available.

With the enormous improvements in the use of computers in engineering analysis (CAE) and design (CAD) many designs are bought to the marketplace without any preliminary models being produced. The manufacturers rely upon the analytical calculations to identify any high-stressed areas that may lead to a design failure. In most cases a mechanical failure in something like a hair drier will only cause an inconvenience to the owner, but when a failure occurs in the aforementioned wing box or fuselage, then the outcome will be a catastrophic accident with a potential for loss of life.

In this instance the manufacturer will produce a full or scale model of the part or assembly and apply appropriate loading to it and monitor any high-stressed areas to see if the part will reach or exceed its calculated design life. The analysis will be undertaken using a variety of instrumentation, including but not restricted to the following:

- Photoelastic coatings
- Brittle lacquer coatings
- Electrical resistance strain gauges

10.1 PHOTOELASTICITY

For many years photoelastic analysis was at the forefront of experimental stress analysis where solid models were manufactured from photoelastic material. These were loaded in a like manner to the actual part, then heated and allowed to carefully cool in a low-temperature oven over several hours. When the model was cured it was carefully cut up and the sections analyzed using a polariscope; the technique relied upon the use of polarized light, and changes in the stress patterns in the material created fringes. These fringes gave an indication of the internal stresses locked into the part, and using special calibration techniques, accurate stresses could be assigned to these fringe patterns. This article discusses the principles behind both three-dimensional and two-dimensional photoelastic models.

10.1.1 THE PRINCIPLES OF PHOTOELASTICITY

The underlying principles will be outlined covering the basic optical instrument, the polariscope, and its different variations are described together with the techniques used for the determination of separating the principal stress values from the photoelastic data.

This chapter will not cover the detailed optical theory of the photoelastic effect and the various forms of the polariscope; this material is given in standard textbooks covering this subject.

Photoelastic stress analysis is an extremely powerful technique that enables a "whole field" analysis to be carried out on a model of the component, compared with using strain gauges, which only consider the strain at a single point on the component.

This work, when carried out at the design stage, will verify the results of a finite element analysis. In some case studies, using the photoelastic method, stresses as much as 20% higher were found than were predicted in the finite element (FE) model. The accuracy of the FE model is totally dependent upon the mesh density. With modern high-speed computers it is possible to increase the mesh density to improve accuracy, but at a cost in computing power.

The technique of photoelastic analysis involves the making of a scale model, either full size or to scale, manufactured from a suitable material (usually Araldite CT 200), and then placing the model in a beam of polarized light using an instrument called a polariscope.

If the model is loaded in a similar direction to that of the actual component, an interference pattern will be visible in the model. The interference pattern will consist of colored fringes called isochromatics. These fringe patterns can be interpreted to give information on the magnitudes of the principal stresses that are present in the model.

Using a simple formula, the stresses in the actual part can then be determined even though the component material may be either steel, aluminum, cast iron, ceramic, or even plastic itself.

A second set of interference fringes is also visible under certain conditions; these fringes are termed isoclinics. Isoclinics enable the directions of the principal stresses at any point in the model to be determined, and from these, stress trajectories or stress flow lines can then be drawn for the actual component under similar loading conditions.

10.1.2 PRINCIPLES

The process is based on the property of birefringence, which is exhibited by certain transparent materials. Birefringence is a property where a ray of light passing through a birefringent material experiences two refractive indices. Many optical crystals exhibit the property of birefringence or double refraction. But photoelastic materials show this property of birefringence only when the model, manufactured from this material, has a stress applied to it and the magnitude of the refractive indices at each point in the model is directly related to the state of stress existing at that point. Thus, the state of the stress in the model will be similar to the state of the stress in the actual structural component.

When a ray of plane-polarized light is passed through a photoelastic material, it will be resolved along the two principal stress directions, and each of these components experiences different refractive indices. The difference in the refractive indices leads to a relative phase retardation between the two component waves. The "stress optic law" gives the magnitude of the relative retardation. where R is the induced retardation, C is the stress optic coefficient, t is the specimen thickness, $\sigma 11$ is the first principal stress, and $\sigma 22$ is the second principal stress.

$$R = CT(\sigma 11 - \sigma 22)$$

The two waves are bought together in a polariscope. The phenomenon of optical interference then takes place and a fringe pattern is produced, depending upon the relative retardation. From the study of the fringe pattern the state of stress at various points in the material can then be determined.

10.1.3 ISOCLINICS AND ISOCHROMATICS

Two types of fringe patterns can be observed in photoelasticity: isoclinic and isochromatic fringes are viewed when the source light is monochromatic. These fringes appear as light and dark fringes; whereas with white light illumination, colored fringes are observed. The difference in principal stresses is related to the birefringence, and hence the fringe color through the stress optic law.

The definition of an isoclinic is that it is the locus of points at which there is a constant inclination of principal stress directions, i.e., wherever either principal stress direction coincides with the axis of polarization of the polarizer.

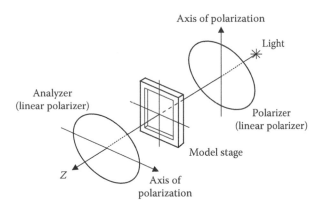

FIGURE 10.1   Schematic of the working components for a plane polariscope.

Isoclinics cannot give information on the magnitude of the principal stress differences, but give valuable information on the principal stress directions.

Isochromatics are the locus of the points along which the difference in the first and second principal stresses will remain the same; hence they are the lines that join the points with equal maximum magnitudes of shear stress. When viewed in white light the fringes are seen in brilliant colored bands and have the same colored sequence as is observed with a film of oil on water and with Newton's rings.

10.1.4 PLANE POLARISCOPE

In this setup the plane polariscope (Figure 10.1) consists of two linear polarizers and a light source. The light source can emit either monochromatic or white light, depending upon the investigation being carried out.

First the light passes through the first polarizer, which converts the light into plane-polarized light. The apparatus is set up in such a way that this plane-polarized light then passes through the stressed test specimen. This light then follows, at each point of the specimen, the direction of principal stress at that point. The light is then made to pass through the analyzer and the fringe pattern can then be observed.

The fringe pattern in a plane polariscope setup consists of both the isoclinics and isochromatics. The isoclinics change with the orientation of the polariscope, while there is no change in the isochromatics.

10.1.5 CIRCULAR POLARISCOPE

Two quarter-wave plates are added to the experimental setup of the plane polariscope. The first quarter-wave plate is placed in between the polarizer and the test specimen, and the second quarter-wave plate is placed between the specimen and the analyzer. The effect of adding the quarter-wave plates is to generate circularly polarized light.

The essential advantage of a circular polariscope (Figure 10.2) over a plane polariscope is that in a circular polariscope setup, only the isochromatics will be seen. The isoclinics are eliminated. This will eliminate the problem of trying to differentiate between the isoclinics and isochromatics.

10.1.6 TWO-DIMENSIONAL AND THREE-DIMENSIONAL PHOTOELASTICITY

Photoelasticity can be applied to either two- or three-dimensional analysis.

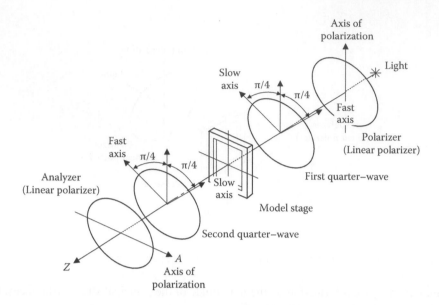

FIGURE 10.2 Schematic of the working components for a circular polariscope.

Two-dimensional photoelasticity is carried out on test specimens manufactured from thin material shaped to the basic profile under investigation, and when mounted in a special test rig, allows loads to be applied to the specimen and the change in the fringes to be observed in real time.

Three-dimensional photoelasticity is more complex. It requires a three-dimensional model to be manufactured and to be fitted in a special heated oven. A steady load is applied to the model prior to a heating program being applied. The model is then heated and cooled under a carefully controlled heating program. This allows stresses generated due to the loading arrangement to be locked into the model on withdrawal from the oven.

This is known as stress freezing technique.

The model is then sliced up in the areas of specific interest. The samples can then be observed and analyzed as two-dimensional specimens using either the plane or the circular polariscopes, depending upon the experimental setup.

10.1.7 FURTHER DEVELOPMENT

Photoelastic analysis has been used on a wide range of stress analyses and was the preferred experimental stress analysis method before the advent of finite element analysis (FEA).

When using FEA, it is crucial to assess the accuracy of the numerical model. This can only be assessed by using experimental verification. As an example, a threaded joint experiences nonuniform contact. This is difficult to incorporate accurately into the finite element model. Idealized models tend to significantly underestimate the actual stress concentrations at the root of the thread.

Photoelastic analysis still remains a major tool in the modern stress analyst's arsenal, validating mathematical models and identifying the accurate placing and orientation of electrical resistance strain gauges.

10.2 PHOTOELASTIC COATINGS

In certain instances it may be desirable to carry out a dynamic or static examination of a physical part, such as a landing gear structure or a gearbox housing to determine the surface strains. With the use of photoelastic coating it is possible to mold a special strain-sensitive coating around the part under examination. The technique is "full field" identical to the two- and three-dimensional methods.

Photoelastic coating has a history of successful applications in a wide range of situations covering manufacturing and construction where stress analysis is used, including aerospace, automotive, pressure vessels, etc. Other applications are found in medical engineering, including the determination of the strength of femur prostheses.

10.2.1 PREPARATION OF THE COATING

A liquid plastic is cast on a flat-plate mold and allowed to partially polymerize. While it is still in a pliable state, the sheet is removed from the mold and formed by hand to the contours of the test part. When fully cured, the plastic coating is bonded in place with special reflective cement and the part is then ready for testing.

For plane surfaces, premanufactured flat sheets are available and can be cut to size and bonded directly to the test part.

10.2.2 ANALYSIS OF THE COATING

The coating is illuminated by polarized light from a reflection polariscope (Figure 10.3). When viewed through the polariscope, the coating displays the strains in identical patterns to the plane and circular polariscopes used on the two-dimensional models.

The polariscope is similar to the circular polariscope (Figure 10.2) and is able to be handheld or mounted on a tripod.

The polariscope can be fitted with an optical transducer (compensator) where quantitative stress analysis can then be quickly performed. The use of photography or video recording can maintain a permanent record of the overall strain distributions, particularly in dynamic analysis.

Although the coating is not permanent, the coating will have a reasonable life for fairly long-term examinations to be carried out. Unlike brittle lacquer coatings, the photoelastic coating method may be used over a wide range of loading spectra. Providing the coating is protected, the results are not subject to environmental effects such as humidity or temperature.

The light source can either be a standard light source for taking static measurements or be replaced with a stroboscopic light accessory for cyclic dynamic measurements.

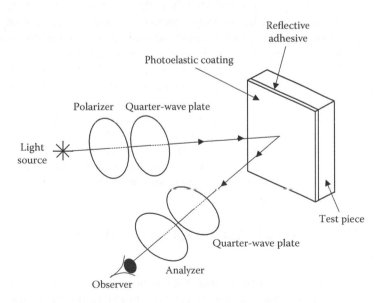

FIGURE 10.3 Schematic representation of a reflection polariscope.

10.2.3 COATING MATERIALS

A wide selection of coatings are available from specialist suppliers in either flat sheet and liquid form for applications to metals, concrete, rubber, and most other materials.

Also available are specially designed application kits for the installation on most test parts.

10.2.4 FULL FIELD INTERPRETATION OF STRAIN DISTRIBUTION

Photoelastic coatings allow strain measurement to be undertaken in either the laboratory or on-site for in situ measurements during construction or manufacture. The advantage of the photoelastic coatings is that it is full field. Areas of high stresses can be easily identified together with the principal stress direction, so that the coating can be locally cut away and an electrical resistance strain gauge fitted, allowing for a quantitative measurement to be taken to confirm the strain reading. It has to be born in mind that the strain gauge reading may not match that of the photoelastic coating; this is due to the length of the gauge chosen. Standard length gauges will have a gauge length of 12.5 mm, and therefore the strain reading will be integrated over this length, resulting in a lower strain reading. If a shorter gauge length is chosen, then the reading will be higher as the integration length is reduced.

The interpretation of the fringe patterns is identical to the two- and three-dimensional photoelastic analysis technique. Photoelastic fringes have their own characteristic patterns, which is very helpful in interpreting the fringe patterns. As an example, the fringes are ordinarily continuous bands, forming either curved lines or closed loops, like the isobars on a weather map. The black zero-order fringes are usually isolated spots, lines, or areas surrounded by higher-order fringes. The fringes never intersect or otherwise lose their identity. The fringe order, and therefore the strain level, is uniform at every point on a fringe. Also, the fringes always exist in a continuous sequence by both number and color; i.e., if the first and third fringes are identifiable, the second-order fringe will lie between them. The color sequence in any direction establishes if the fringe order and strain level are increasing or decreasing in that direction.

If there is a zero-order fringe within the field of view, it will be obvious by the black color. Assuming that the coated test piece has a free square corner or pointed projection, the stress in this area will always be zero, and a zero-order fringe will exist at this position, irrespective of the load magnitude. The zero-order fringe may shrink slightly as the load magnitude is increased.

From Table 10.1 note the fringe order counting from the zero-order fringe. It is important to note that the analyst should have good color vision; otherwise misinterpretation of the color fringes will occur. Most people tend to have a problem with the color green, and this could lead to reading the stresses a magnitude higher. To overcome this problem, it is possible to view the model in monochromatic light and physically count the fringes from the zero-order fringe to the area of interest.

Generally, when viewing the photoelastic coating without any load applied, it is possible to note the direction of the fringes as the load is incrementally applied. Without any load being applied the color of the coating will be dominated by the zero-order fringe (black), and as the load is progressively applied, additional colored fringes in the correct color sequence will begin to appear and spread across the test piece. The analyst will note the appearance of the fringes; closely spaced fine fringes will denote steep strain gradients. Areas where almost uniformly black or gray exist usually indicate a significantly understressed region.

10.3 INTRODUCTION TO BRITTLE LACQUER COATINGS

10.3.1 INTRODUCTION

The advantages offered for using brittle lacquer is it can be applied to a wide range of materials, including metals, many plastics, glass, ceramics, and wood products. It is inexpensive and requires no instrumentation to analyze the results.

TABLE 10.1
Isochromatic Fringe Characteristics

Color	Fringe Order
Black	0.00
Pale yellow	0.60
Dull red	0.90
Red/blue transition	1.00
Blue-green	1.22
Yellow	1.39
Rose red	1.82
Red/green transition	2.00
Green	2.35
Yellow	2.50
Red	2.65
Red/green transition	3.00
Green	3.10

The coating is sprayed on to the surface and then allowed to cure in either air or heat to attain the brittle properties. As the component is loaded, the coating will begin to crack when the threshold strain or strain sensitivity is exceeded.

10.3.2 LOADING AND TESTING TECHNIQUES

The brittle lacquer coating fractures in response to the surface strain underneath it. Analysis of the coating indicates the direction and magnitude of stress within the elastic limit of the test piece material. The brittle lacquer coating gives a graphic picture of the distribution, direction, location, sequence, and magnitude of the tensile strains. The coating cracking is a predetermined value, which is determined by a simple calibration procedure.

Under normal conditions the coating will crack in the range from 5.08×10^{-6} to 76.2×10^{-6} m/m. In terms of stress these values represent between 41 and 620 MPa.

The qualitative picture of the stress distribution gives an immediate visual indication for immediate design improvements. With very careful handling it is possible to obtain quantitative results with a ±10% accuracy.

Flaking of the brittle lacquer coating is an indication of local yielding in the test piece.

Cracking occurs where the strain is highest, giving an immediate indication of the presence of stress concentrations. The cracks also indicate the directions of maximum strain at these areas, as they always align at right angles to the direction of the maximum principal tensile strain. The method has been in use for many years, principally to provide test engineers with quick and reliable information about the strain response of the material to which the coating has been bonded. In addition, as it provides information on the principal strain directions, this information is of great use in the selection, location, and orientation of strain gauges for the accurate measurement of peak stresses.

10.3.3 EFFECTS OF CHANGE IN RELATIVE HUMIDITY AND TEMPERATURE

The lacquers generally have a nominal threshold sensitivity of 500 microstrain at specified temperatures and humidity conditions specified by the lacquer manufacture in a coating selection chart. The selection of a suitable coating depends upon the environmental conditions occurring at the time of the test.

The coating begins to lose its brittleness and sensitivity begins to shift as the relative humidity rises above 20%. This only begins to become noticeable over large changes in the relative humidity.

Other disadvantages include that it is affected by humidity and temperature, i.e., requiring the test temperature to be maintained so that an approximate strain level can be determined by comparison with calibration test pieces.

Strain sensitivity changes by approximately 4 microstrain per 1% change in relative humidity.

The effect of temperature on the coating is approximately 70 microstrain per degree Celcius change. As the temperature increases the coating begins to soften and becomes more plastic, and therefore loses its sensitivity. The opposite applies as the temperature falls. According to StressKote (a major supplier of brittle lacquer coatings), a simple ready calculation is that a 5°C change will cause a sensitivity shift to 360 microstrain, whereas at a constant temperature a 25% change in relative humidity will cause a shift of 100 microstrain.

The grade of coating selected for the test will be dependent upon the test temperature and relative humidity.

Prior to the test program, a calibration test strip is prepared and loaded as per the manufacturer's instructions. The calibration is carried out at the test temperature and relative humidity. This allows the microstrain sensitivity of the coating to be read from the calibration test strip.

The calibration bar is simply a rectangular bar and fits into a calibration jig that supports the bar at one end. The bar is then essentially a cantilever. An eccentric cam is located at the free end of the cantilever, and this produces a known strain distribution along the cantilever length. When the lacquer on the bar is fully cured, the cam is rotated to depress the bar by a known amount. The position at which the cracking of the lacquer begins gives the strain sensitivity when compared with a marked strain scale. Quantitative measurements of strain levels can then be made on the component under test, as, for example, if the calibration sensitivity is shown to be 1,000 microstrain (strain × 10^{-6}), then the strain at the point on the component at which cracks first appear will also be 1,000 microstrain.

10.3.4 MEASURING STRAIN UNDER STATIC LOADING

The technique required is to incrementally load the component, holding the load for a few minutes and releasing it to zero prior to the application of the next increment, noting the crack patterns at each load application. It is recommended to mark key information with a wax pencil. Photographs can be taken with a camera fitted with a flash. Care should be taken when using floodlights, for if they are too close to the test piece, the heat may soften the coating and cause the cracks to heal.

10.4 INTRODUCTION TO STRAIN GAUGES

The most widely used experimental stress analysis technique in industry today is the strain gauge. The gauges come in different types, but the most common are vibrating wire strain gauges and electrical resistance strain gauges.

10.4.1 VIBRATING WIRE STRAIN GAUGES

Vibrating wire stain gauges are mostly used in civil engineering for monitoring bridge and tunnel loads. They generally have a large gauge length and are not unduly affected by the size of the aggregate used in concrete. The gauge comes in a variety of fittings. It may be embedded in concrete, able to be spot welded to steel beams or columns among a variety of alternative attachments.

The basic construction is that of a fine wire supported under tension between two supports (see Figure 10.4). As the structure to which the gauge is secured moves, this will cause tension on the wire and in turn affect the frequency of the wire. When a measurement is required, the

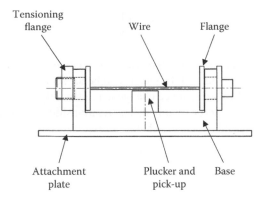

FIGURE 10.4   Schematic diagram of a vibrating wire strain gauge.

wire is electrically "plucked" and a pick-up then measures the frequency of the vibrating wire. Instrumentation converts this frequency into a strain reading.

The advantage of the vibrating wire gauge is that it is able to lie dormant for several hours or days without any power being applied to it, and when required it can be switched at will to give an accurate reading without any further calibration required. Indeed, the connection to the gauge can be severed, and provided the gauge position is not disturbed, a calibrated signal will be available when the signal cable is reconnected. Other advantages include being weatherproof and able to operate in any environmental conditions, including corrosive atmospheres provided adequate protection is provided.

Because of its large gauge length (it can be supplied in shorter gauge lengths) the vibrating wire strain gauge has little application in engineering measurements outside of the field of civil engineering.

10.4.2   Electrical Resistance Strain Gauges

In 1843 Charles Wheatstone published his work on the bridge circuit he had invented. He had noticed the change in resistance in an electrical conductor due to the effects of mechanical stress.

William Thomson (1824–1905) went further with some work published in 1956.

In 1938, two workers in the United States, Edward E. Simmons and Arthur C. Ruge, developed the basics of the modern strain gauge.

The electrical resistance strain gauge is simply a length of wire or foil formed into a shape of a continuous grid, as shown in Figure 10.5. This grid is cemented to a nonconductive backing.

The gauge can then be bonded securely to the surface of the component under investigation. Any strain experienced in the surface of the component will be transmitted to the gauge itself.

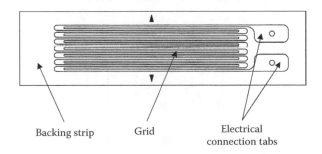

FIGURE 10.5   Elements of a linear electrical resistance strain gauge.

The fundamental equation for the electrical resistance R of a length of the conductor is

$$R = \frac{\rho L}{A} \tag{10.2}$$

where L is the length of the conductor, A is the cross-sectional area, and ρ is the specific resistance or resistivity. Any change in the length of the gauge, and hence sectional area, will result in a change of resistance. Measurement of this resistance change using suitably calibrated equipment enables a direct reading of the linear strain to be obtained. The relationship that exists for a number of alloys over a substantial strain range between the change in resistance and strain may be expressed as follows:

$$\frac{\Delta R}{R} = K \times \frac{\Delta L}{L} \tag{10.3}$$

where ΔL and ΔR are the change in length and resistance, respectively, with K being termed the gauge factor.

Thus,

$$\text{Gauge factor } K = \frac{\Delta R/R}{\Delta L/L} = \frac{\Delta R/R}{\varepsilon} \tag{10.4}$$

where ε is the strain. The strain gauge manufacturer will always supply the value of the gauge factor and can be checked using simple calibration procedures if required. Typical values of K for most conventional foil strain gauges will lie in the region of 2.0 and 2.2, and most modern strain gauge instruments, with the value of K set accordingly, allow the strain values to be recorded directly.

Changes in resistance produced by normal strain levels that are experienced in engineering components are very small, and as a consequence, sensitive instrumentation is required to measure these changes. Strain gauge instruments are based on the Wheatstone bridge networks as shown in Figure 10.6. The condition of balance for this network is

$$R_1 \times R_3 = R_2 \times R_4 \tag{10.5}$$

In its simplest half-bridge wiring system, gauge R_1 is the active gauge, i.e., the gauge actually being strained. Gauge R_2 is called a dummy gauge and is bonded to a piece of material identical to that which R_1 is bonded to, but in this case it is unstrained. The purpose for this is to cancel out in gauge R_1 any resistance change due to temperature changes. Gauges R_1 and R_2 represent the working half of the network—hence the name half-bridge system (Figure 10.7). Gauges R_3 and R_4 are standard resistors built into the measuring instrument. Alternative wiring systems utilize one (quarter bridge) or all four (full bridge) (Figure 10.8) resistors of the bridge resistance arms.

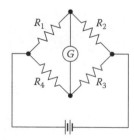

FIGURE 10.6 Wheatstone bridge circuit.

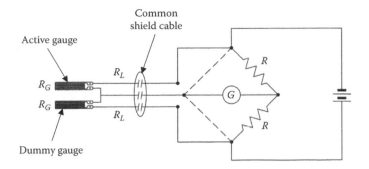

FIGURE 10.7   Three lead-wire system for half-bridge (dummy-active) setup.

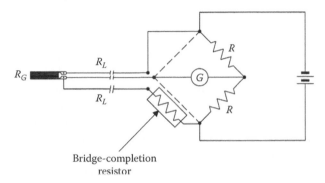

FIGURE 10.8   Three lead-wire system for quarter-bridge (dummy-active) set-up with a single self-temperature compensated gauge.

10.4.3   UNBALANCED BRIDGE CIRCUIT

With the bridge initially balanced to zero, any strain that occurs at gauge R_1 will cause the galvanometer needle to deflect. The deflection can be calibrated to read the strain in R_1, as stated above, and the circuit can have an arrangement where the gauge factor can be adjusted. Strain readings can therefore be taken with the pointer off the zero position and the bridge is thus unbalanced.

10.4.4   NULL BALANCE OR BALANCED BRIDGE CIRCUIT

An alternative measurement procedure is to make use of a variable resistance in one arm of the bridge. This will cancel out any deflection of the galvanometer needle. This adjustment can be calibrated directly as strain, and the readings are taken with the pointer on zero, i.e., in the balanced position.

10.4.5   INSTALLATION PROCEDURES

The quality and success of any strain gauge installation are influenced by the consideration and choice of the installation procedure together with the choice of the adhesive. The apparently mundane procedure of actually cementing the gauge onto the component is a critical step in the operation. Every precaution must be taken to ensure a scrupulously clean surface if perfect adhesion is to be attained. J. Pople covers full details of the surface preparation and installation techniques in the BSSM (British Society for Strain Measurement) *Strain Measurement Reference Book*.

Following the successful attachment of the gauge to the surface, lead wires are then attached to the electrical connection tabs using solder. Extreme care is required in this operation, as it is very easy to generate a dry joint with its associated noise. An insulation check needs to be carried out to confirm the installation; the installation resistance to ground (RTG) should be a minimum of 104 megaohms at 30 V.

The sequence for successfully installing an electrical strain gauge is as follows:

1. Determine the measurement required. What feature do you want to measure? Bending, torsion, tension, compression, etc.?
2. Select the most appropriate strain gauge for that measurement.
3. What is the specimen material? Select the appropriate adhesive for the particular application and compatible with the gauge backing.
4. Select the area where the gauge is being fitted and ensure there is adequate access for fitting the gauge.
5. Follow the installation procedure carefully, ensuring the gauge, if clamped, is equal over its length and the pressure is between 34 and 138 kPa. Unequal clamping will result in an unequal glue line.
6. Select the style of terminal pad and bond adjacent to the strain gauge.
7. Select the correct type of cable between the gauge and bridge. This is of vital importance, as the cable in a quarter-bridge configuration forms part of the measuring circuit. Any change in the cable resistance will affect the accuracy of the strain measurement.
8. Select the correct type of solder and flux for the application: only use the solder and flux that has been approved by the gauge manufacturer. Many types of commercial solders and fluxes are unsuitable for strain gauge applications.
9. Carefully solder the jumper leads between the strain gauge and the terminal pad and the lead wire from the terminal pad to the bridge.
10. The lead wire may have to terminate in a plug and socket at an appropriate point if the measuring point is a distance from the instrumentation. It is essential that the resistance remains consistent during the measurement phase and the resistance value is as low as possible.
11. Depending upon the application, a choice will have to be made if the strain gauge requires protection. This will depend upon the environmental conditions the gauge is expected to operate in. If the gauge is operating in clean air conditions within laboratory conditions and the test is expected to last a very short time, the gauge can in this circumstance be painted with a paint compatible with the lead insulation to prevent any oxidation at the gauge-structure interface. But if the test is being undertaken in workshop conditions, then a minimal degree of protection will be required. Encapsulating the gauge in an insulating layer of polysulfide rubber (Bostik 2114-5) will provide this. This will give protection for up to 6 months against any moisture ingress.
12. In situations where the gauge is expected to operate in the most adverse conditions, a suitable metal surround will be required, particularly if the gauge has to face sea water conditions such as on oil rigs, together with the extremes of temperature. Layers of polysulfide epoxy further protect the polysulfide rubber encapsulation. The mechanical housing has an extension to it providing a totally leak-proof connection for the signal cable. This is also fitted with an adequate insulation.
13. At several points in the installation of the gauge, the insulation will have to be tested using a gauge insulation tester to ensure that the insulation resistance is in excess of 20,000 megaohms for foil strain gauges when installed under laboratory conditions. A value of 10,000 megaohms should be considered a minimum. Readings below this figure generally indicate trapped foreign matter, moisture, residual flux, or backing damage due to soldering, as well as incomplete solvent evaporation from the overcoating.

10.5   EXTENSOMETERS

Extensometers were developed for use on tensile testing machines for measuring strains; they measure the very small extensions or contractions in the test pieces when under increasing load. They are not precluded from use in measuring strains in other applications. The extensometer was invented by Dr. Charles Huston in the 1870s (Figure 10.9).

Modern extensometers utilize electronics in their construction, using a very sensitive displacement transducer. Special clip-on fasteners enable the extensometer to be quickly attached to the test specimen. These instruments are extremely accurate, having a resolution of less than 0.3 μm. These instruments require a conditioning module for their operation.

These extensometers are known as *contact extensometers*.

10.5.1   CONTACT EXTENSOMETERS

Figure 10.10 shows an extensometer for measuring strains in plastic test pieces.

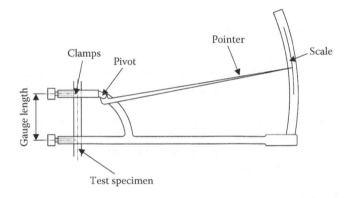

FIGURE 10.9   Huston's extensometer.

FIGURE 10.10   Extensometer for plastic test coupons.

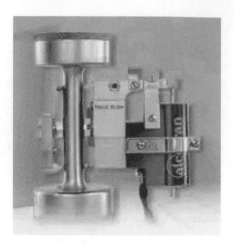

FIGURE 10.11   Extensometer for metal test coupons.

10.5.2   NONCONTACT EXTENSOMETERS

Another class of extensometers, known as *noncontact*, utilize lasers for performing the strain or elongation measurements. The principle utilizes the reflection of a laser light from the surface of the test specimen via a charge-coupled interface digital (CCD) camera. Resolutions of typically 0.1 μm can be attained.

Targets are attached to the test specimen at a known distance apart, and a laser unit scans the targets and measures the distance apart. Although costly compared to the contact type, they are very easy to set up. The laser light is very low power, and therefore, providing the operator does not stare into the light, they are safe to use.

10.5.2.1   General Notes

The big disadvantage with these types of extensometers is the "gauge length," which tends to be rather large for very local strain measurements. As with the electrical resistance strain gauge as well as the vibrating wire gauges, the strain is integrated over a larger length than those of the electrical resistance strain gauges, and hence are very local strains or not identified. Having said that, they are on a par with the vibrating wire strain gauges, where these do have quite large gauge lengths.

In civil engineering, where most of the structures are made using concrete, the size of the aggregate will dictate the gauge length. In some instances the size of the aggregate can be as large as 25 mm, in which case the gauge length will need to be greater than 100 mm to be able to yield any meaningful results.

The disadvantage with these, as with the electrical resistance strain gauges, is that the datum reading will be lost if there is a power failure. Therefore it is essential to have a reliable battery backup.

10.5.3   APPLICATIONS

As with the vibrating wire strain gauges, these extensometers have found a use in the civil engineering and mining industries for monitoring walls, etc. Plotting displacements against time, geotechnical engineers can predict if a failure is imminent.

11 Introduction to Fatigue and Fracture

11.1 INTRODUCTION AND BACKGROUND TO THE HISTORY OF FATIGUE

The earliest investigation into metallic fatigue was undertaken by August Wöhler between 1852 and 1870, who had been commissioned by the French Railway Company SNCF. It had experienced a number of axle failures and wanted to know and understand the reasons for them.

Wöhler constructed a rotating bending test rig (see Figure 11.1) to emulate and investigate the loading on a full size railway axle. These axles were subjected to a range of rotation speeds and loads.

Figure 11.2 shows a representation of the rotating cantilever test rig identifying its major component part.

He constructed a curve (Figure 11.3) comparing the applied load on the axle to the number of cycles to failure. From this curve he observed that the curve exhibited an endurance limit where stresses below that limit did not display any further damage.

From these results it was established that metals had a limiting stress where failure could be estimated with a reasonable degree of accuracy under the conditions prevailing at that time.

The next breakthrough in understanding the mechanics of fatigue was observed by Bauschinger in 1886. His experiments showed that the limit of elasticity on the first compression loading is less than the initial tension loading; this is called the Bauschinger effect. Figure 11.4 shows that if point C is extrapolated, C′, using the B-C slope, then $\Delta\varepsilon_b$ will show the permanent cyclic strain softening or cyclic strain hardening. The second part of Figure 11.4 is drawn to an absolute scale for stress–strain.

When strain hardening occurs, as in cast aluminum, this effect becomes more apparent. These plots are generated during a fatigue test showing the Bauschinger effect and are often referred to as hysteresis loops.

The first fatigue tests were conducted using a fully reversible cycle, i.e., alternating loads between compression and tensile with the mean stress at 0.

This situation very rarely occurs in real life, and a way was found to offset the normal Wöhler type curve. Goodman, Soderberg, and Gerber undertook this work. Their work is based on either the ultimate tensile stress σ_{uts} or yield stress σ_{yield}, and the endurance limit stress σ_e or the fatigue strength for a given number of cycles σ_{am} (for zero mean stress) as the baseline for a safe design and is drawn on two axes, with the x axis representing the mean stress and the alternating stress plotted on the y axis. Figure 11.5 shows these three most widely used empirical relations. The straight line joining the alternating fatigue stress to the tensile stress is the modified Goodman law. Goodman's original law included the assumption that the alternating fatigue limit was equal to one-third of the tensile stress, but this has been modified to the relation shown, using the alternating fatigue stress determined experimentally. The original law is no longer used and has been replaced by the modified law; this is referred to as the Goodman law.

Gerber deduced that the early results found by Wöhler fitted closely to a parabolic relation, and this is often referred to as Gerber's parabola. A straight line connecting the alternating fatigue stress to the static yield stress gives the third relation, known as Soderberg's law. For many purposes it is essential that the yield stress is not exceeded, and this relation is intended fulfill the condition that neither fatigue failure nor yielding shall occur.

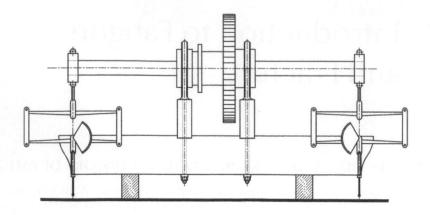

FIGURE 11.1   Wöhler's original railway axle.

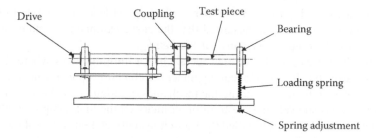

FIGURE 11.2   Representation of a rotating cantilever test rig.

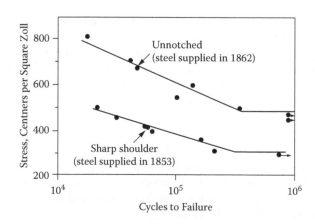

FIGURE 11.3   The S-N curve as described by Wöhler. (1 centners per zoll² = 0.75 Mpa.)

The Goodman curve gives good results for brittle materials and conservative results for ductile materials. The Gerber relation will give good results for ductile materials.

It was found that a compressive mean stress improved the fatigue life of a component and decreased with a tensile mean stress. Gerber's parabolic relationship may therefore produce erroneous results to the conservative side in the compressive mean stress region.

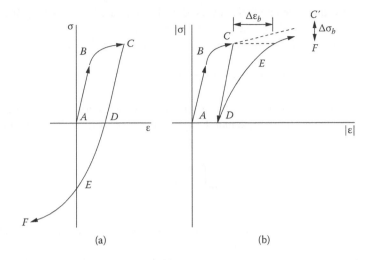

FIGURE 11.4 The stress and strain curve for cyclic loading.

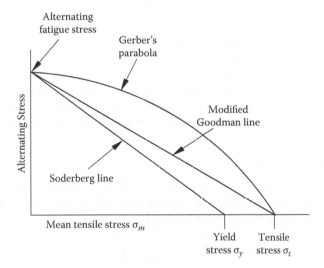

FIGURE 11.5 Fatigue stress–static stress diagram.

11.1.1 LATER DEVELOPMENTS

Old notions regarding the fatigue mechanism based on the crystallization theory were proved incorrect by Ewing and Rosenheim in 1900, closely followed by Ewing and Humfrey in 1903, showing that slip bands developed in many grains of the polycrystalline material. These slip bands broaden as cyclic deformation continues, leading to extrusions and intrusions on the surface of the component, and then one dominant flaw will then lead to failure.

The empirical relationship of the stress-cycle or S-N curve was proposed by Basquin in 1910 using a log-log scale showing the linear relationship (see Equation 11.1).

$$\frac{\Delta\sigma}{2} = \sigma_a \sigma_f' \cdot \left(2 \cdot N_f\right)^b \qquad (11.1)$$

where $\Delta\sigma$ is the stress range for one fully reversed cycle, σ_f is the fatigue strength coefficient, b is the fatigue stress exponent, and N_f is one fully reversed cycle.

One fully reversed cycle will be in the following sequence (see Figure 11.4(a)):

$$A \rightarrow B \rightarrow C \rightarrow D \rightarrow E \rightarrow F \text{ back to A}$$

Palmgren in 1924 and Miner in 1945 also noticed the accumulation of damage and presented a linear cumulative damage rule that excluded any sequence of loading. The equation is shown in Equation 11.2:

$$\sum_{i=1}^{m}\left(\frac{n_f}{N_f}\right) = 1 \tag{11.2}$$

where n_i is the number of cycles corresponding to the amplitude σ_{ai}, which is the stress amplitude, and N_{fi} is the number of cycles to failure σ_{ai}.

Notch effects at the tip of a notch had been found by Neuber to follow the well-known rule as in Equation 11.3.

$$K_t = \sqrt{K_\sigma \cdot K_\varepsilon} \tag{11.3}$$

where K_σ is the ratio of the maximum local stress to the nominal stress, and K_ε is the ratio of the maximum local strain to the nominal strain.

With strain being the concentration now, this will also address the plastic deformation at the tip of the notch. The same can also be performed for the fatigue notch factor by replacing K_t with K_f.

K_t can be viewed as the macroscopic stress-strain notch factor, and K_f as the stress-strain notch factor associated with the microscopic properties, such as material, surface finish, and inclusions.

11.1.2 RECENT DEVELOPMENTS

Further development was independently undertaken by Coffin and Manson, where they proposed that plastic strain was responsible for fatigue damage. They also proposed an empirical relationship between the number of load reversals to failure and plastic strain.

$$\frac{\Delta\varepsilon_p}{2} = \varepsilon_f' \cdot \left(2 \cdot N_f\right)^c \tag{11.4}$$

where $\Delta\varepsilon p$ is the strain range for one fully reversed cycle, ε_f' can be set equal to ε_f, which is the fracture ductility in a simple monotonic test, and c is in the range of -0.5 to -0.7 for most metals.

Fracture mechanics has its origins in the stress analysis of Inglis and the energy method of Griffith. In their analysis of brittle solids and later work by Irwin it was shown that the amplitude of the stress singularity ahead of the crack tip can be expressed by the stress intensity factor, K.

The stress fields at the crack tip can be derived for three major modes of loading, each involving different crack surface displacements (Figure 11.6):

Mode I: Opening or tensile mode. Crack surfaces moving directly apart.
Mode II: Sliding or in plane shear mode. Crack surfaces moving relative to one another, normal to the crack front.
Mode III: Tearing or antiplane shear mode. Crack surfaces moving relative to one another parallel to the crack front.

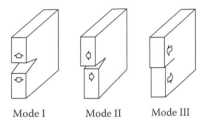

Mode I   Mode II   Mode III

FIGURE 11.6   Various modes of failure.

Most engineering situations involving cracked components are covered by mode I loading. Mode III occurs in pure shear situations. Mode II is rarely encountered except in rolling contact bearing races. Mixed modes are encountered in biaxial loading situations.

Equation 11.5 gives a relationship for the stress intensity factor:

$$K_1 = F \cdot \sigma \cdot \sqrt{\pi \cdot a} \tag{11.5}$$

with I generally indicating mode I, Fa geometrical factor that will refer to the crack size and loading condition, all in relation to the specimen, and "a" the crack size in depth. Further work by Paris, Gromez, and Anderson showed that the range in stress intensity factor could be related to the change in fatigue crack length per load cycle. Equation 11.6 shows this law.

$$\frac{da}{dN} = C \cdot \Delta K^m \tag{11.6}$$

where:

$$\Delta K = K_{max} - K_{min}$$

where C and m are material constants.

The knowledge of crack initiation and subsequent slip bands that formed to initiation cracks has been increased with the development and use of the electron and optical microscopes. This has led to the discovery of so-called persistence slip bands (PSBs) by Zappfe and Worden. The term *persistence* refers to the phenomenon that even after a layer of surface material is removed where the slip bands have previously occurred, the slip bands will occur in the same place.

The next area of interest is that the stress intensity factor can change as a result of the crack advancement. In other words, the ΔK can be influenced by the details of crack closure or fracture surface contact in the wake of the advancing fatigue crack tip. Elber showed that under certain conditions cracks will stay closed in a cyclic tensile load. The fatigue crack growth is therefore not only influenced by the values of ΔK alone, but also by the prior history and crack size. The so-called short crack phenomenon was identified by Pearson, who showed that small fatigue flaws, typically an order smaller than other fatigue cracks, will exhibit faster crack growth to when the crack lengthens. Figure 11.7 shows a typical crack growth rate curve.

This has been a brief introduction into the mechanics of fatigue and fracture. The following sections in this chapter will give more detail on this subject.

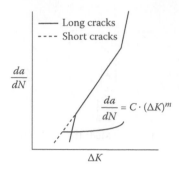

FIGURE 11.7   Crack-growth rate curve.

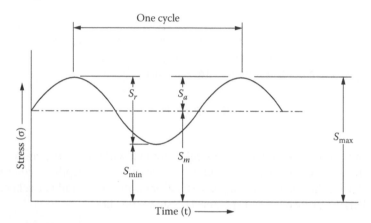

FIGURE 11.8   Stress versus time curve.

11.1.3   BASIC DEFINITIONS

The following are definitions of some terms frequently used in the discussion of fatigue analysis (please refer to Figure 11.8):

Stress cycle: The smallest division of the stress-time function that is repeated.

Nominal stress: Obtained from the simple theory in tension, bending, and torsion, neglecting geometric discontinuities.

Maximum stress: The largest or highest algebraic value of a stress in a stress cycle. Positive for tension (S_{max}).

Minimum stress: The smallest or lowest algebraic value of a stress cycle. Positive for tension (S_{min}).

Mean stress: The algebraic mean of the maximum and minimum stress in one cycle (S_{mean}).

Stress range: The algebraic difference between the maximum and minimum stresses in one cycle (S_r).

Stress amplitude: Half the value of the algebraic difference between the maximum and minimum stresses in one cycle, or half the value of the stress range (S_a).

Stress ratio: The ratio of minimum stress to maximum stress.

Fatigue life: The number of stress cycles that can be sustained for a given test condition.

Fatigue strength: The greatest number of stress cycles that can be sustained by a member for a given number of stress cycles without fracture.

Fatigue limit: The highest stress level that a member can withstand for an infinite number of load cycles without failure.

Fatigue life for P percent survival, N_p: The fatigue life for which P percent of the sample has a longer life; for example, N90 is the fatigue life for which 90% will be expected to survive and 10% will fail.

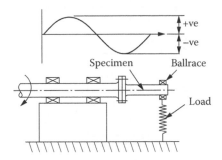

FIGURE 11.9   Schematic arrangement of a Wöhler fatigue test.

11.2   THE FATIGUE PROCESS

From the introduction it is seen that from the tests undertaken by Wöhler, the catastrophic failures of the rail axles were due to fatigue. The axles had failed due to the repeated application of a sinusoidal cyclic load, each of a magnitude, well below the yield stress of the material used. Clearly it can be seen that at any point on the axle, it would be subjected to a fully reversed stress cycle each single revolution of the axle. See Figure 11.9 for a schematic arrangement.

Modern fatigue testing very rarely subjects the complete component or assembly to a fatigue test, but usually relies on subjecting a standardized test specimen to the test loads. This has the advantage that the test is subjected to identically similar geometric components, and a detailed database can be developed comparing the results loading, frequency, and materials.

Modern fatigue testing subjects the standardized specimens to a wide range of types of testing strategies, including push-pull, two- and three-point bending, fluctuating torsion, and other modes of stressing besides rotating bending.

The nature of the testing has to be carefully considered in light of the nature of the failure mode being investigated. In Wöhler's case he subjected his test pieces to a rotating bending test. It would not be appropriate to compare the results from a push-pull test with those of a rotating bending test. They are essentially totally dissimilar tests.

The simple rotating cantilever test imposes a sinusoidal loading cycle about a zero mean stress. More complex fatigue tests may superimpose a fluctuating stress on a mean stress that will not be zero or use load cycles that may not be sinusoidal.

In carrying out fatigue tests, a batch of nominally identical test specimens is prepared and a single specimen is subject to a load W and the number of cycle to failure N_f is counted and logged. The remaining specimens are tested at other loads, W_1, W_2, W_3, ...,W_n, to determine the lives N_{f1}, N_{f2}, N_{f3}, ..., N_{fn}.

Stress amplitudes S corresponding to the full load cycle are calculated according to the specimen size and geometry, and the results are presented as a plot of S vs. log N_f. A typical plot is shown diagrammatically in Figure 11.10, and these graphs are referred to as S/N curves.

There are two types of S/N curves:

a. This curve shows a definite horizontal portion or "knee" at long lives.
b. Here the curve becomes asymptotic to the N axis at very high values of N.

For curve a, the horizontal portion of the curve defines a value of S below which failure is not likely to occur, however large the value of N becomes. This value of S is known as the fatigue limit of the material (or endurance limit) when expressed in units of stress. This type of behavior is typical of ferrite steels (and also some aluminum alloys). For most materials this value, as a rough guide, is about 40% of the uniaxial tensile strength.

In contrast, most nonferrous alloys and polymeric materials exhibit an S-N curve, which continues to fall as the stress amplitude is lowered. Curve b in Figure 11.10 is a typical example, and the

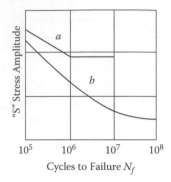

FIGURE 11.10   Typical fatigue curves.

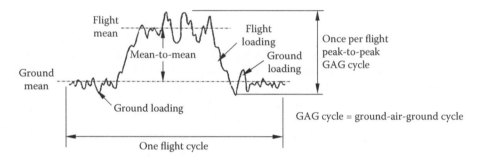

FIGURE 11.11   Example of a typical flight cycle.

endurance limit is defined as the stress amplitude necessary to cause fatigue failure after a specified number of stress cycles (for example, $N = 10^7$ cycles).

Fatigue testing is not limited to the rotating bending mode. Other tests include push-pull, plane bending, fluctuating torsion, and other modes of stressing. The control of load amplitude may be replaced by other parameters, such as total strain, plastic strain amplitude, or displacement.

The simple rotating cantilever test imposes a sinusoidal load cycle acting about a zero mean stress on the specimen. More complex tests may superimpose a fluctuating stress on a mean stress that is not zero, or use a load cycle that is not sinusoidal and employs load sequences that attempt to simulate service conditions (Figure 11.11). At its most complex, a whole structure such as a vehicle chassis or aircraft structure may be subjected to computer-controlled cyclic loading that more approximately represents those that will be experienced in service.

Fatigue testing is essential for the acquisition of data to help predict a fatigue failure in a component. This is performed in a variety of ways, depending upon the stage of the design or production phase. The following main types of tests can be identified:

- Stress-life testing of small specimens
- Strain-life testing of small specimens
- Crack growth testing
- S-N tests of components
- Prototype testing for design validation

The first three tests are generally idealized tests using standard specimens; these are used for producing information on the material response. The use of the results from these tests in life prediction of components and structures requires additional knowledge of the influencing factors related to size, geometry, surface conditions, and any corrosive environment.

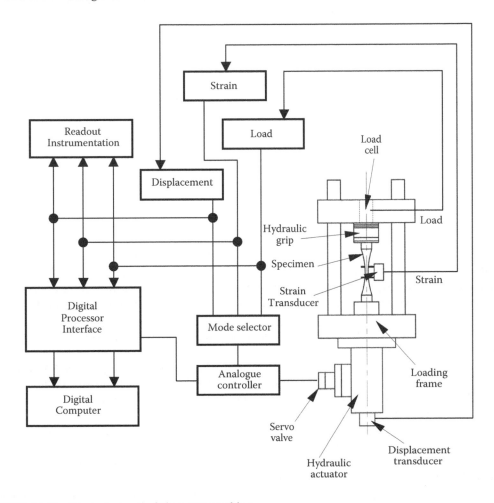

FIGURE 11.12   Electrohydraulic fatigue test machine.

Life prediction tests are normally carried out using standardized specimens for producing more accurate S-N curves that are not influenced by the presence of notches and other surface conditions.

Rotating bending machines have been used in the past for generating large amounts of test data in a relatively inexpensive way. These are now being replaced by the use of closed-loop electrohydraulic test machines; these are fitted with hydraulic grips that facilitate the insertion and removal of the test specimens.

Figure 11.12 depicts an axial (push-pull) electrohydraulic servo fatigue test machine that is widely used in the majority of material test laboratories. These machines are capable of precise control of almost any type of stress-time, strain-time, or any load patterns.

Special equipment such as environmental chambers allow the test specimens to be tested at high or low temperatures in addition to providing corrosive atmospheres around the specimen and evaluate the fatigue characteristics in these atmospheres.

11.3   INITIATION OF FATIGUE CRACKS

Fatigue failure in metals starts with the initiation of a crack invariably at a free surface. Then follows a period where the crack begins to nucleate with other local cracks and starts a slow growth until the remaining section cannot carry the load and a catastrophic separation occurs.

There are two distinct fatigue processes involved: initiation and propagation.

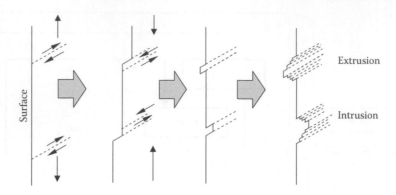

FIGURE 11.13 Wood's model for fatigue crack initiation.

The initial crack in engineering alloys will often occur at a site other than a slip band. Grain boundaries, second-phase particles, particle-matrix interfaces, corrosion pits, and machining marks have all been observed at sites of crack initiation. The common feature of these sites is that they must be capable of localizing plastic deformation that occurs in directions that intersect a free surface.

If a crack or crack-like defect preexists, then there will be no initiation process required and the whole life of the specimen is then spent in crack growth.

The cyclic slip process (shear controlled) has been widely held to be responsible for crack initiation on the surface of uniform polished specimens. It is believed that slip will occur between crystals orientated in the direction of the maximum shear stresses, which is roughly at an angle of 45° to the applied cyclic load.

Several equivalent models have been proposed to try and explain the initiation of fatigue cracks by local deformation. Wood has proposed a model depicted in Figure 11.13. During the rising load part of the cycle, slip occurs on a favorably orientated slip plane. In the falling load part slip takes place in the reverse direction on a parallel slip plane.

Since slip on the first plane is inhibited, strain hardening is by oxidation of the newly created free surface. This first cyclic slip can give rise to an extrusion or intrusion in the metal surface. An intrusion can grow into a crack by continuing plastic flow during subsequent cycles.

An intrusion-induced crack or imperfection appears on the metal surface, and once started (stage I crack growth) can also grow by a mechanism of reversed slip, leading to microcracking, which in turn will lead to a macrocrack (stage II crack growth).

Many components live most of their lives with microcracks preexisting within them. It is obvious that for many components it is the ability to propagate cracks rather than initiate them that is important.

Stage I initiation (slip band growth) may be completely absent in practical cases where cracks originate from highly stressed, sharply notched. Stage I growth can account for up to 90% of the total fatigue life of smooth ductile specimens at low stresses.

Stage II propagation is the later stages of the crack growth and represents the remaining 10% of the growth to failure. The fracture surfaces may show the well-known macroscopic progression marks known as beach marks.

These marks are often curved, with the center of curvature at the origin. They serve as a useful guide to direct the investigator to the fracture initiation site.

11.4 FACTORS AFFECTING FATIGUE LIFE

It would be perfect if the effect of any single variable on the separate processes of stage I and stage II crack growth was fully understood. Unfortunately, many investigators have only studied the effects of particular variables on the total number of cycles to failure (N_f). The effect is further complicated, as a single variable may have different effects in low-stress/high-cycle fatigue from those of high-stress/low-cycle fatigue. Variables may also interact; as an example, frequency of cycling is apparently a single variable, but altering it will usually also change strain rate and the time available for environmental effects at the crack tip opened by the tensile component of the load cycle.

11.4.1 STRESS AMPLITUDE

The effect of stress amplitude on the cycles to failure is large, as indicated by the S/log N curves, and it is generally accepted that the tensile component is the most damaging. The effect of periodic large amplitude loading is complex, and it may be either beneficial or damaging. A large load early in the life may induce some work hardening or blunt a notch or groove by plastic deformation; either event will have beneficial results. A tensile overload during stage II growth may induce compressive residual stresses at the crack tip, slowing down crack growth during the immediate or immediately subsequent cycles. A sufficiently large tensile overload during stage II growth may produce a burst of overload failure that will extend the crack a considerable distance. This effect is referred to as crack jumping.

Summarizing, increasing the stress amplitude will result in early crack initiation and an increase in the rate of both modes of crack growth.

11.4.2 MEAN STRESS

It is unusual for engineering components to be subject to stress cycling about a zero mean stress. It is more usual for an alternating stress to be superimposed on a mean stress that is either compressive or tensile (see Figure 11.14).

In the case where the stress amplitude is expressed as a maximum and minimum, as in Figure 11.15, the mean stress is conveniently expressed by a dimensionless parameter R, where:

$$R = \frac{\sigma_{min}}{\sigma_{max}}$$

It is well established that tensile mean stress reduces fatigue life, and plotting the stress amplitude S against mean stress shows the relationship σ_{mean} for a particular value of N_f (see Figure 11.16).

As described in Section 11.1, a number of models have been proposed for the analytical and graphical representations of the relationship between the mean life and the alternating stress for

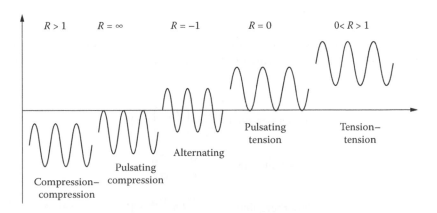

FIGURE 11.14   Stress cycles with different mean stresses and "R"-ratios.

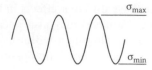

FIGURE 11.15 Amplitudes of the stress cycle.

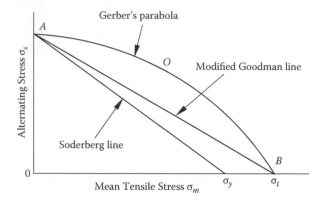

FIGURE 11.16 Comparison of Gerber, Soderberg, and Goodman laws viewed against stress amplitude and mean tensile stress.

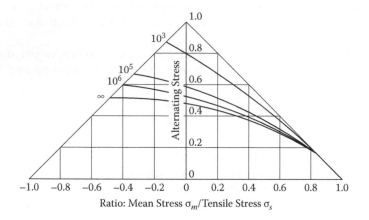

FIGURE 11.17 Typical master diagram.

a given fatigue life. The most well-known among these models are those proposed by Gerber, Goodman, and Soderberg. The life plots are displayed in Figure 11.17 and are described by the following expressions:

$$\text{Gerber parabola:} \quad \frac{\sigma_a}{\sigma_e} + \left(\frac{\sigma_m}{\sigma_{ult}}\right)^2 = 1$$

$$\text{Goodman line:} \quad \frac{\sigma_a}{\sigma_e} + \frac{\sigma_m}{\sigma_{ult}} = 1$$

$$\text{Soderberg line:} \quad \frac{\sigma_a}{\sigma_e} + \frac{\sigma_m}{\sigma_{ys}} = 1$$

where:

σ_a = Fatigue strength in terms of stress amplitude when $\sigma_m \neq 0$

σ_e = Fatigue strength (for a fixed life) in terms of stress amplitude for a fully reversed loading ($\sigma_m = 0$ and $R = -1$) in some references $\sigma_e = \sigma_{ao}$)

σ_{ult} = Ultimate tensile strength of material

σ_{ys} = Material yield strength

Most metals and alloys give results that lie between the Goodman's line and the Gerber's parabola, with the Goodman's line matching experimental data quite closely for brittle materials but is conservative for ductile alloys.

The Soderberg line is more conservative than the Goodman line but is physically more meaningful in that it puts the maximum possible mean stress as equal to the yield stress.

Gerber's parabola is generally good for ductile alloys.

Whole families of these diagrams are needed to describe the behavior of a material over a range of N_f values, and it is simpler to show the effect of R on a so-called master diagram of the sort shown in Figure 11.17. It is usual in such diagrams to normalize alternating stress (plotted on the y axis) and the mean stress (plotted on the x axis) by dividing the values by the tensile stress. A single diagram can thus be used as a guide to a whole range, such as steel or aluminum alloys.

Such diagrams may be plotted from experimental data but are frequently derived from empirical formulas. Heywood uses the following formula to plot a master diagram giving the fatigue life of unnotched steels as a function of mean stress and alternating stress amplitude:

$$\frac{\sigma_a}{\sigma_t} = \left(1 - \frac{\sigma_m}{\sigma_t}\right)\left[A_o + \gamma(1 - A_o)\right]$$

where

$$A_o = \frac{1 + 0.0038N^4}{1 + 0.008N^4}$$

and

$$\gamma = \frac{\sigma_m\left(2 + \dfrac{\sigma_m}{\sigma_t}\right)}{3\sigma_t}$$

where:

σ_m = mean stress

σ_a = alternating stress

σ_t = ultimate tensile stress

N = cycles to failure

11.5   STRESS CONCENTRATIONS

Any discussion on fatigue analysis has to include a discussion on the effects stress concentrations have on the life of a component.

Consider a component in Figure 11.18 manufactured with a hole in the middle of the strip and subject to a tensile force P.

At position 1, the stress is uniform across the section and equal to the load divided by the cross-sectional area, i.e.:

$$\sigma_1 = \frac{P}{A_1}$$

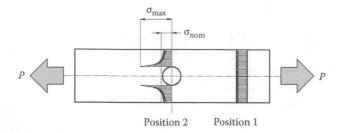

FIGURE 11.18 Effect of a stress concentration.

At position 2, it is clearly seen that the stress is significantly higher at the edge of the hole and then drops off rapidly with the transverse distance from the hole. At a distance of about one hole diameter, the stress is less than P/A_2; this is due to reasons of equilibrium. That is:

$$\int_{A_2} \sigma_2 dA + P$$

(11.8)

The integral of the normal stress over the whole area must equal the applied force, since the average normal stress times the total area balances the applied force. Therefore, if the stress is greater than P/A_2 on one portion of the cross section, it has to be less than P/A_2 elsewhere.

11.5.1 THE ELASTIC STRESS CONCENTRATION FACTOR

The index of the severity of the stress concentration is referred to as the stress concentration factor, and the commonly accepted symbol is K_t for statically loaded parts.

The stress concentration factor is defined as the ratio of the maximum stress to the nominal stress at the section containing the discontinuity, or as in Figure 11.18.

$$K_t = \frac{\sigma_{2max}}{\sigma_{2nom}} = \frac{\sigma_{2max}}{\dfrac{P}{A_2}}$$

(11.9)

By way of an example:

If $P = 20{,}000$ N and $A_2 = 120$ mm^2,

$$\sigma_{2nom} = 167 \text{ MPa}$$

In this instance, if the maximum stress was found to be 420 MPa,

$$\sigma_{2max} = 420 \text{ MPa}$$

Then

$$K_t = \frac{420 \text{ MPa}}{167 \text{ MPa}} = 2.51$$

In this instance a safety factor as high as 2, based on the nominal stress at the critical section, would still be insufficient to prevent a failure. This is a case where a safety factor should not be used to allow for the presence of a stress concentration.

11.5.2 THE FATIGUE STRESS CONCENTRATION FACTOR

The presence of any surface discontinuities, either external or internal, as discussed previously, will have a significant effect on the endurance limit of the component.

The static stress concentration factor will not be applicable to the stress concentration factor used for fatigue analysis. The stress concentration factor for fatigue is denoted K_f. This is influenced by the presence of notches from which, due to the high stress concentration, cracks may start to grow, depending upon the severity of the notch. The effect of the notches in fatigue K_f is defined as the unnotched to notched fatigue strength, obtained from fatigue tests (see Figure 11.19).

$$Kf = \frac{\text{fatigue strength of unnotched specimen}}{\text{fatigue strength of notched specimen}}$$

Both material strengths are measured at the same value of N_f.

The difference in magnitude of K_f and K_t for some materials permits the definition of a notch sensitivity factor q where

$$q = \frac{K_f - 1}{K_t - 1} \tag{11.10}$$

For fully notch-sensitive materials $K_f = K_t$ and $q = 1$. For totally notch-insensitive materials $K_f = 1$ and $q = 0$. Ductile materials have lower notch sensitivity than stronger, more brittle materials, so that little benefit is usually gained in changing from a low-strength steel in attempting to improve notch fatigue life. Heywood gives excellent and detailed accounts of notch fatigue strength and extensive data on many materials.

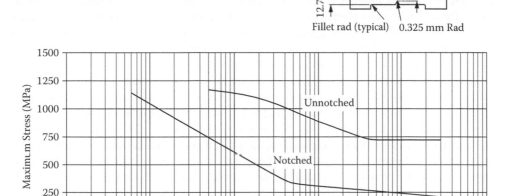

FIGURE 11.19 Experimental S-N curves for a notched and un-notched specimen.

11.6   STRUCTURAL LIFE ESTIMATIONS

Estimation of fatigue life is a fairly simple process but one beset with several complexities. Obviously load magnitude and sequence are very important elements of the process. Although a number of techniques have been developed to satisfy specific conditions, the simplest and most widely used, as well as practical, is the Palmgren-Miner hypothesis. As discussed in Section 11.1, Palmgren devised the technique in 1924 relative to calculating the life of ball bearings. In 1945 Miner presented a paper reporting its application to structural elements. The Palmgren-Miner method merely proposes that the fraction of the fatigue life used up in service is the ratio of the applied number of load cycles at a given level divided by the allowable number of load cycles to failure at the same variable stress level. If several levels of variable stresses are applied to a part, then the sum of the respective cycle ratios is the fraction of fatigue life used up.

This damage is usually referred to as the cycle ratio or cumulative damage ratio. If the repeated loads are continued at the same level until failure occurs, the cycle ratio will be equal to unity. When fatigue loading involves a number of stress amplitudes, the total damage is the sum of the different cycle ratios and failure should still occur when the cycle ratio sum equals unity.

where:

$$D = \sum_{i=1}^{k}\left(\frac{n_i}{N_i}\right) = 1 \qquad (11.11)$$

where
n_i = Number of loading cycles at the ith stress level
N_i = Number of loading cycles to failure for the ith stress level based on a constant amplitude
k = Number of stress levels to be considered in the analysis

Fatigue crack initiation is assumed to occur when the D value is equal to unity. There are three parameters, which affect the magnitude of the summation of the cycle ratios.

1. The order of the load applications. Consider, for example, two different stress levels, f_1 and f_2, and their respective cycle lives, N_1 and N_2. Consider also that f_1 is greater than f_2; if f_1 is applied prior to f_2 the life will be shorter than if f_2 is applied first.
2. Summation of cycle ratios. The second effect on the summation of cycle ratios is due to the amount of damage caused by continuous loading at the same level. The summation of cycle ratios for different stress levels is accurate only if the number of continuous cycles at each stress level is small. For most applications, the loading is random and the stress level is constantly changing. The number of continuous cycles at a particular level should be small and the summation of cycle ratios is considered fairly accurate.
3. Notched and unnotched features. This parameter affects the summation of the cycle ratios and whether or not the fatigued part is notched (such as fastener holes, etc.) or unnotched. The unnotched part generally gives a summation less than unity, while the notched part gives a summation greater than unity. Since most of the structural fatigue failures originate in some form or other from a notch, it indicates that a good average value of 1.5 should be used for the cycle ratio to predict failure of, say, an airframe major structural member, such as a wing box. For simpler structural members a cycle ratio of 1.0 should be used.

11.7   INTRODUCTION TO LINEAR ELASTIC FRACTURE MECHANICS

Notation:
a   = Crack half-length for a crack free to extend at both ends
a_c = Critical crack length at failure

a_0 = Initial crack length
2_b = Characteristic dimension of component width
C = Coefficient in simple crack growth rate law defined by $da/dN = C(\Delta K)^n$
E = Modulus of elasticity
G = Release rate of potential energy/unit area
K = Stress intensity factor (mode I)
K_c = Critical value of stress intensity factor (also plane stress fracture toughness)
K_{Ic} = Plane strain fracture toughness in mode I
ΔK = Stress intensity range
N = Number of cycles
R = Stress ratio, S_{min}/S_{max}
r = Polar coordinate of stress in component
S_c = Critical stress at failure
t = Thickness of component in the region of the crack tip
α = Geometric factor in stress intensity factor
σ_y = Yield (or 0.2% proof) stress of material
σ_r = Stress along axis of polar coordinate r
θ = Angle of polar coordinate r from the x axis
ν = Poisson's ratio

11.7.1 PREAMBLE

The first major study into the fracture phenomenon in cracked bodies was undertaken by A.A. Griffith, who presented the paper "The Phenomena of Rupture and Flow in Solids" in 1921 following his work on glass specimens. His basic premise was that unstable propagation of a crack occurs if any increment of crack growth results in more stored energy being released than being absorbed by the creation of the new crack surface. From further testing he developed a constant failure value relating to the applied stress field and crack length over a range of crack lengths.

During World War II there were a series of catastrophic failures involving mass-produced Liberty cargo ships that fractured into two pieces on or immediately after being launched. There followed further spectacular failures immediately after the war, notably the Comet aircraft in the 1950s, and there was a resurgence of activity in developing the concept of fatigue and failure analysis. G.R. Irwin evolved an alternative interpretation of fracture phenomena known as the stress intensity factor approach in the 1950s. This work then focused attention on the mechanics near the crack tip and is now widely used for solving both the residual strength and fatigue life calculations.

11.7.2 COMPARISON OF FATIGUE AND FRACTURE MECHANICS

Similarities and dissimilarities between fatigue and fracture mechanics are summarized in Table 11.1. Both fatigue and fracture mechanics rely on results from testing; however, the fracture mechanics concept makes it possible to handle fracture considerations in a quantitative manner and has shown greater applicability to fatigue crack propagation.

11.7.3 THE DIFFERENCE BETWEEN CLASSICAL FATIGUE ANALYSIS AND FRACTURE MECHANICS

The fundamental difference between fatigue analysis and fracture mechanics is that in fatigue analysis, the analysis is based on the rate of crack growth. No consideration is given to the preexistence of flaws or cracks that would eventually lead to the generation of a starting crack, or the time spent in the stage I phase. The S-N curves indicate the average number of cycles taken to a failure for a given stress value, the specimens either precracked or being manufactured with a standard groove in the test coupon. This groove creates a stress concentration that then leads to the starting crack in

TABLE 11.1
Similarities and Dissimilarities between Fatigue and Fracture Mechanics

Fatigue Characteristics or Considerations	Fracture Mechanics Characteristics or Considerations
Considers no initial material flaws, e.g., voids, inclusions, etc.	Assumes pre-existence of flaws, inhomogeneities and discontinuities in a material.
Data presented in the form of a plot of stress versus number of cycles to failure, S.N. curve.	Data presented in the form of stress intensity factor versus cycles to failure or flaw growth rates.
Life prediction utilises cumulative damage theories.	Life prediction is based on minimum flaw growth potential, i.e., the growth of an initial flaw to critical value.
Analysis carried out in two steps:	
1. Relating repeated loads to stress.	Imposes limits on non-destructive inspections and procedures.
2. Evaluating stresses using the cumulative damage theory to predict structural life.	Predicts fatigue behaviour such as those stemming from stress corrosion or fatigue.
Does not consider sustained loading.	
A purely analytical fatigue design method is not yet available.	Considers sustained loading.
	Considers sequence of operational load.
The scatter inherent in fatigue behaviour and in service conditions would require that results be interpreted statistically.	Has shown greater applicability to fatigue crack propagation because conditions for fatigue less than critical.
Considers fractures for relatively large numbers of cycles only (10.10^3 and over).	Considers fractures for relative small numbers of cycles ($0 < $ cycles $< 10.10^3$)

a small number of cycles. The original fatigue test results for a large number of materials, including aluminum alloys and steels, have been lost; therefore it is not possible to subject these results to any modern statistical analysis.

The underlying premise of fracture mechanics is that all parts contain minute sharp-edge cracks. These cracks grow in a stable manner under either static or cyclic stress, and when a crack has reached a critical length, the growth rate then becomes unstable, leading to fast or brittle fracture on the next load cycle.

Referring to cracks as defects does not imply that the material is defective; it implies that it is inevitable that engineering size cracks and other discontinuities such as voids, cold-shuts, and incomplete weld penetrations will exist.

11.7.4 STRESS INTENSITY

Fracture mechanics gives a numerical description of the elastic stress field near the crack tip, called the stress intensity factor, and characterizes the behavior of the stress field just beyond the small plastic zone at the crack tip in the presence of small-scale yielding.

The stress intensity factor relates the level and mode of the loading, as well as geometry, to the stress state at the crack tip. It has to be emphasized that the factor is not to be considered as a stress concentration factor in the geometric sense of the term. However, having said that, the square root of the crack length used in the stress intensity factor corresponds to the square root of the notch radius used to calculate the stress concentration factors.

There is no single equation that describes the general relationship between the applied stress and stress intensity. The relationship for a specific case is dependent on the mode of loading and geometry.

For a constant stress amplitude and through thickness cracks, the stress intensity factor is given by

$$K_1 = \sigma\sqrt{\pi a} \qquad (11.12)$$

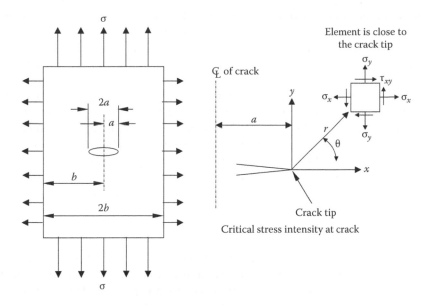

FIGURE 11.20   Mode I crack under a biaxial stress field.

11.7.4.1   General Stress Intensity Solution

Irwin used classical stress analysis methods to investigate the detailed stress distributions near the crack tip. Based on the complex stress function approach of Wesergaard, Irwin showed that the elastic stress field in the neighborhood of the crack tip (see Figure 11.20) was given by

$$\sigma_x = \frac{K}{\sqrt{2\pi \cdot r}} \cos\left(\frac{\theta}{2}\right)\left[1 - \sin\left(\frac{\theta}{2}\right)\sin\left(\frac{3\theta}{2}\right)\right] \tag{11.13}$$

$$\sigma_y = \frac{K}{\sqrt{2\pi r}} \cos\left(\frac{\theta}{2}\right)\left[1 + \sin\left(\frac{\theta}{2}\right)\sin\left(\frac{3\theta}{2}\right)\right] \tag{11.14}$$

$$\tau_{xy} = \frac{K}{\sqrt{2\pi r}} \cos\left(\frac{\theta}{2}\right)\sin\left(\frac{\theta}{2}\right)\sin\left(\frac{3\theta}{2}\right) \tag{11.15}$$

It should be noted that these stress distributions are inversely proportional to the square root of the distance from the crack tip. At the crack tip itself (r = 0), the stress distributions predict infinite stresses, but this is an idealized situation, known as a stress singularity, resulting from the assumption of elastic behavior without any limiting criterion. On the plane of the crack ($\theta = 0$, y = 0), the shear stress is zero and the direct stress components are given by

$$\sigma_x = \sigma_y = \frac{K}{\sqrt{2\pi r}} \tag{11.16}$$

The term $\sigma\sqrt{(\pi r)}$ is dependent only on the applied stress and crack size and defines the gradient of stress with inverse square root of the distance away from the singularity at the crack tip. The term $\sigma\sqrt{(\pi r)}$ was defined by Irwin as the stress intensity factor and given the symbol K. It should be noted that K is not a stress concentration factor and has dimensions with units of stress x $\sqrt{(\text{length})}$.

Although the definition of the stress intensity factor as $K = \sigma\sqrt{(\pi r)}$ is the one generally used, in the case of a central crack in an infinite plate subject to remote tension, there are some papers in the literature where an alternative definition is used without the π, namely, $K = \sigma\sqrt{(a)}$. Care must be taken to check which definition is being used in any particular case. In these notes the Irwin definition of $K = \sigma\sqrt{(\pi r)}$ is used throughout.

It is important to recognize that the stress singularity and the stress intensity factor, which dominate the stress field at the crack tip, are features of tension loading. These arise because tension forces cannot be transferred across the free surface of the crack and are redistributed around the edges of the crack in a nonuniform manner. When compression loading is applied to the cracked plate, if the crack surfaces are in contact, forces can be transmitted directly through the crack, so there is no requirement for redistribution, and hence no stress singularity, and the stress intensity factor is zero. This has important consequences when fatigue loading is applied to a cracked component.

11.7.5 FRACTURE TOUGHNESS AND CRACK GROWTH

Fracture toughness is the second premise after the stress intensity factor of fracture mechanics, and it characterizes the crack growth of a material containing a defect, either surface breaking or subterrarium, and is a material property found by experiment (usually performed using the Charpy test).

Fracture toughness is the ability of a part containing a crack or defect to sustain a load without catastrophic failure. As long as the stress intensity factor K stays below a critical value of K_{Ic} (the parameter representing a critical value for fracture toughness in mode I) (see Figure 11.6), the crack is considered stable. If K reaches or exceeds K_{Ic}, the crack will propagate and lead to sudden failure. Propagation rates can reach speeds over 1,500 m/s.

Analytical techniques involving the application of fracture toughness data have become indispensable tools for the design of fail-safe structures, particularly those involving high-strength materials.

From the work of Paris and Erdogan it has been established that there is a simple relationship between crack growth rate and the range of stress intensity factor during the loading cycle.

$$\frac{da}{dN} = A(\Delta K)^m \tag{11.17}$$

This is known as the Paris-Erdogan law, where A and m are constants and determined experimentally.

If the crack propagation law for the material is known, it is possible to calculate by integration the number of cycles required for the crack to propagate from one length to another. Further, if the fracture toughness value for the material is known, then it is possible with the value of the maximum design stress to calculate the critical value of crack length at which fast fracture will occur—hence, by integration of the Paris law, the total life in cycles of the cracked component.

During fatigue crack growth, $\Delta K = \Delta\sigma a^{1/2}.\alpha$, where α is the compliance factor for the given geometry. The crack growth law gives $da/dN = A(\Delta K)^m = A(\Delta\sigma.\alpha c^{1/2})^m$.

Integrating this expression for the number of cycles required for the crack to grow from an initial length a_0 to the critical length for fast fracture gives:

$$\int_{a_i}^{a_c} a^{(-m/2)}da = A\alpha^m \Delta\sigma^m \int_{N=i}^{N=N_f} dN$$

$$= \frac{2}{2-m}\left[a_i^{(1-m/2)} - a_f^{(1-m/2)}\right] = A\alpha^m \Delta\sigma^m N_f$$

Note that this expression fails at m = 2.

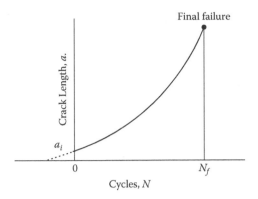

FIGURE 11.21 Fatigue crack as a function of life measured in cycles of stress.

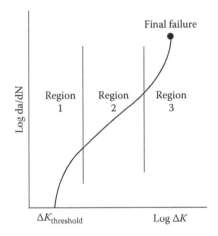

FIGURE 11.22 Fatigue crack growth rate as a function of stress intensity function.

Early literature reported values of m mostly lying between 2 and 4, although values much higher have been encountered. It is becoming clear that for the purpose of the above calculation appropriate values of m lie between 2 and 3, the higher values being found in materials of low toughness. Thus, when m = 3,

$$N_f = \frac{2}{A^3 (\Delta\sigma)^3} \left(\frac{1}{\sqrt{a_i}} - \frac{1}{\sqrt{a_f}} \right) \tag{11.18}$$

Although integrating for total life is a useful procedure, more information is obtained if the calculation is carried out in a stepwise manner. This demonstrates that growth of the crack accelerates with respect to life, as shown in Figure 11.21.

In fact, the real situation is worse than this because it is now known that the Paris-Erdogan law applies only over the middle range of crack growth rates, i.e., between rates of about 10^{-6} and 10^{-4} mm/cycle. A plot of log da/dN against log ΔK shows three regimes of behavior, and the central regime in which the Paris-Erdogan law applies is preceded and followed by the regimes in which m varies and takes much higher values (Figure 11.22).

Region 1 shows that there exists a value of ΔK below which the crack is nonpropagating; i.e., it merely opens and closes without growing forward. This is called the threshold for fatigue crack growth, ΔK_{th}. The rate of crack growth in the threshold region is much slower than calculation from the Paris-Erdogan law would predict.

If ΔK can be held below the value for ΔK_{TH} for the specific material, then an effective infinite fatigue life can be obtained despite the presence of cracks. This does tend to lead to very low design stresses even for small macrocracks.

In region 3, as K_{max} approaches the limiting fracture toughness of the material, K_{IC} or K_C, the Paris-Erdogan law underestimates the fatigue crack propagation rate. This acceleration of the logarithmic growth rate is associated with the presence of noncontinuum fracture modes, such as cleavage, intergranular, and fibrous fractures (these are activated at high levels of K). It is also found there will be a marked sensitivity to the mean stress.

11.8   FATIGUE DESIGN PHILOSOPHY

When designing a part or structure that will be subject to fatigue loading, consideration needs to be given to the design philosophy concept that the assembly has been designed against. There are two design philosophy concepts currently being used: fail-safe and safe-life.

11.8.1   Fail-Safe

A structure designed for a fail-safe philosophy will support designated loads with any single member failed or partially damaged. Sufficient stiffness shall remain to prevent divergence, sever vibrations, or other uncontrolled conditions within the normal design envelope.

Summarizing the fail-safe philosophy:

- Structure has the capability to contain fatigue or other types of damage
- Requires knowledge of:
  - Multiplicity of structural members
  - Load transfer capability between members
  - Tear-resistant material properties
  - Slow crack propagation properties
- Inspection controls
- Fatigue is a maintenance problem

11.8.2   Safe-Life

Safe-life components are those whose failure would result in a catastrophic failure with the potential for loss of life. These components must remain crack-free during their service life. The life of the component or structure in cycles is obtained by fatigue analysis or fatigue testing results divided by an appropriate life reduction factor, which will indicate that premature fracture is extremely remote.

The design philosophy will dictate the level of inspection that a component or structure is subjected to and the time when the part is withdrawn from service.

Summarizing the safe-life philosophy:

- Structure resists damage effects of variable load environment
- Requires knowledge of:
  - Environment
  - Fatigue performance
  - Fatigue damage accumulation
- Limit to service life
- Fatigue is a safety problem

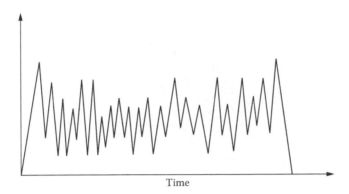

Time

FIGURE 11.23   Schematic stress history of a detail subject to random variable amplitude loading.

11.9   CYCLE COUNTING METHODS

11.9.1   INTRODUCTION TO SPECTRUM CYCLE COUNTING

The actual service history of an aircraft, automotive, or rail vehicle is variable amplitude spectrum waveforms. To predict the durability of a structure, it is essential to decompose the complex variable cycles of the history data into events that can be processed with constant amplitude fatigue data, which are normally obtained from test or laboratory results. The procedure converts a complex irregular load-time history (Figure 11.23) into a series of simple peak-trough events containing individual cycles, equivalent to constant amplitude cycles. These are characterized by mean and amplitude values (or max/min values), called cycle counting.

Various methods for cycle counting of a load sequence can be found from literature. The most widely used counting methods are range-pair counting and the rain-flow counting. Both of these methods define individual cycles as closed stress-strain hysteresis loops.

There are a number of methods used for cycle counting; these include peak stress, level crossing (with or without reset levels), and range-mean. These methods have been shown to overestimate the fatigue life of a component subject to cumulative fatigue. Range-mean is found to most closely approach the true situation but is strongly influenced by the appropriate gate level or threshold used.

The rain-flow technique approaches closest to the true loading sequence, and in nearly all cases will give the most conservative fatigue life estimate.

It is this method that will be described here in more detail. The rain-flow or Pagoda roof method works in an analogy with raindrops falling and rolling on successive roofs. When one looks at the load history turned through 90°, with the time axis now as the vertical axis (Figure 11.25), it is clearly seen why it is called a Pagoda roof. This method requires one to keep track of all the wet parts of the roof.

Figure 11.24 shows the load history of a component that will be analyzed using the rain-flow method. The rain-flow rules follow:

1. Rain-flow starts at the beginning of the test and again at the inside of every peak.
2. Rain flows down a pagoda roof and over the edge, where it falls vertically until it reaches a level opposite a maximum more positive (minimum more negative) than the maximum (minimum) from where it started.
3. Rain also stops when it is joined by rain from the pagoda roof above.
4. The beginning of the sequence is a minimum if the initial straining is in tension.
5. The horizontal length of each rain-flow is then counted as a half-cycle at the strain range.

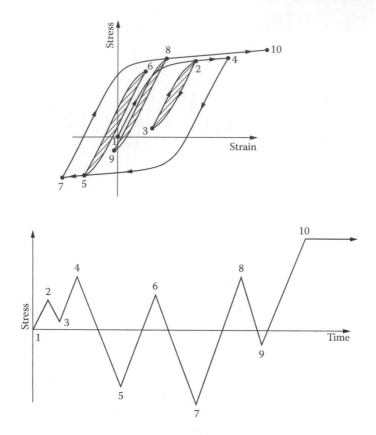

FIGURE 11.24 Stress–strain response for strain history shown in Figure 11.25.

In Figure 11.25 the rain-flow starts at 1, which is considered a minimum. It continues down to 2 and from there to 2′, and then down to 4, and finally stops opposite of 5. This is a (minimum) peak more negative than 1. Thus, a half-cycle from 1 to 4 is extracted.

The next rain-flow starts inside peak 2 and proceeds to 3, falling to a point opposite 4, which is more positive than the starting maximum 2. A half-cycle 2-3 is therefore extracted. A third flow starts at 3 but terminates at 2′, due to the rainfall from above 2. The half-cycle 3-2′ is extracted and paired with 2-3, since 3-2′ and 2-3 form a closed stress-strain loop 2′-3-2.

The next rain-flow begins inside peak 4 and proceeds to 5. The range 1-4 is now counted as a half-cycle. The rain-flow continues down to 5′ and then down to 7, where it falls to a point opposite of 10. This is a maximum greater than the maximum from which the rainfall started at 4. A half-cycle 4-5-7 is then extracted.

Rain now begins to flow from inside peak 5 and then continues down to 6, where it falls to a level opposite of 7, this being a greater minimum than 5. A half-cycle 5-6 is then extracted.

The flow from inside of peak 6 ends at 5′, which is where rain from above joins it. The half-cycle 6-5′ is paired with 5-6 to form a complete cycle 5′-6-5, which is extracted.

The next peak to be considered is 7, from where rain proceeds to 8 down to 8′ and then on to the final peak at 10. The range 4-5-7 is now counted as a half-cycle. The half-cycle 7-8-10 is then extracted. Rain starting from inside of 8 continues to 9 and then downwards until it is level with 10, which is more positive than 8. A half-cycle 8-9 is extracted and paired with the final half-cycle 9-8′. A full cycle 8′-9-8 is then counted.

Thus the strain-time record shown contains three full cycles, 2′-3-2, 5′-6-5, and 8′-9-8, plus three half-cycles or reversals, 1-2-4, 4-5-7, and 7-8-10. Figure 11.25 shows that this combination of cycles

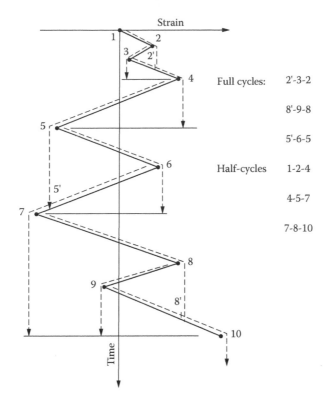

Full cycles: 2'-3-2

8'-9-8

5'-6-5

Half-cycles   1-2-4

4-5-7

7-8-10

FIGURE 11.25   Rainflow diagram.

and half-cycles is a true representation of the stress-strain behavior of the material exposed to the strain-time history. The three closed loops are clearly seen, as are the three half-cycles.

Note that each part of the strain-time record is counted once and once only. It is assumed that damage caused by a large event is not affected by its interruption to complete a small stress-strain loop. The damage of the interruption is simply added to that of the larger cycle or half-cycle.

Using this method a varied amplitude history can be reduced to a series of half-cycles whose maximum can be calculated and the cycles or half-cycles completely defined. Fatigue life can be calculated from constant amplitude data using a cumulative damage law.

12 Introduction to Geared Systems

12.1 INTRODUCTION

Rotary transmission between shafts is accomplished using different methods:

1. Pulleys, using v-belts, timing belts, chains
2. Hydraulic pumps and motors, servo-electro motors
3. Gearing

It is this third group that will be discussed in this chapter. Very simplistically, a gearbox is a device for mechanically transmitting power from one shaft to another and either maintaining, increasing, or decreasing the speed N of the second shaft to the first. Because of the speed change there will be a change in the torque T.

The gearbox designer's prime task is to design a system, which will provide the required characteristics to match the demands of the drive system, whether it be a metal-cutting machine tool drive, automotive gearbox, winch drive, etc.

Gears are wheels that mesh with each other through interlocking teeth. Rotation of one wheel will cause the rotation of the other, albeit in the opposite direction, without any slip between them. There are a number of gear teeth designs; the most common is the cycloidal and involute form. The fundamental basis of the involute form is there is no sliding between the tooth surfaces, the action is purely rolling, and hence there is minimum damage between the teeth, providing the tooth size is adequate for the power being transmitted and there is adequate lubrication in the gearbox.

12.2 TYPES OF GEARS

There are many different types of gears and the classification or identification depends largely on how the gears are used. The most common combination is where the two shafts are parallel with each other; here the gears will be either spur or helical and the shafts will rotate in the opposite direction to each other. When the two shafts are not parallel with each other the connecting gears can be either skew or spiral or even bevel gears.

This chapter will identify the more common configurations.

12.2.1 Spur Gears

This type of gear (Figure 12.1) is most commonly used to connect two parallel shafts that rotate in opposite directions. The teeth are parallel to the axis of the gear and are identical in profile. There is no axial force generated due to the tooth loading, and therefore used as sliding gears for change speed mechanisms in gearboxes.

The spur pinion (i.e., the gear having the smaller number of teeth in a pair of mating gears) is also used to mesh with an internal spur gear. The principal problem with straight-cut spur gears is that they are inherently noisy.

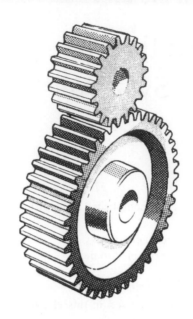

FIGURE 12.1   Spur gear.

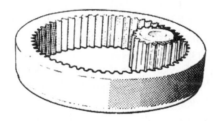

FIGURE 12.2   Internal spur gear.

12.2.2   INTERNAL SPUR GEARS

In this case the teeth are cut on the internal diameter of an annulus or ring (Figure 12.2). The teeth are parallel to the axis and the same profile meshing in a similar manner to the spur gear. The gears are used to connect two parallel shafts usually opposed, and when engaged, the shafts run in the same direction.

This type of gear is used in epicyclic transmissions, reduction gears for machine tools, and for clutches and couplings. In the latter case, however, the external spur pinion has the same number of teeth as the internal wheel and therefore rotates as a single body in the same direction.

12.2.3   RACK AND PINION

The pitch line of a spur rack moves in a longitudinal direction and the rack may be considered a spur gear of infinite radius or zero curvature (Figure 12.3).

A spur pinion (as in Figure 12.1) is most commonly used to transmit the longitudinal motion to the rack, and the most common application can be found on lathes, when in this case the rack is used to transmit motion to the saddle. The rack tooth is of fundamental importance to the study of involute teeth.

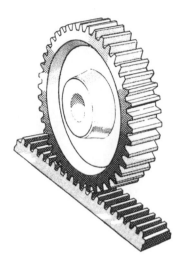

FIGURE 12.3   Rack and pinion.

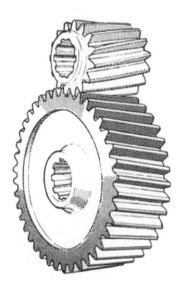

FIGURE 12.4   Helical gears.

12.2.4   HELICAL GEARS

Helical gears (Figure 12.4) are used in the same way as spur gears, but differ from them in that the teeth are cut on the periphery of the gear blank and are of helical or screw form.

The pinion cannot be moved longitudinally along its axis without imparting angular motion to its mating wheel, and for this reason helical gears are used in automotive gearboxes as constant mesh on syncro-mesh applications.

The transverse section of a helical gear is identical with a spur gear insofar as the tooth profile is concerned and is involute on this section. Mating gears must, however, be made with the same helix angle but opposite hand, and the tooth loading produces an axial thrust.

Helical gears are considered quieter than spur gears, and where noise will be a problem, these will be used in those applications where possible.

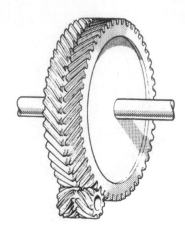

FIGURE 12.5   Double helical gears

FIGURE 12.6   Spiral bevel gear.

12.2.5   DOUBLE HELICAL GEARS

As the name implies, double helical gears (Figure 12.5) are the equivalent of two helical gears secured together, but in actual practice they are manufactured as one piece. The teeth may be continuous or separated by a gap, although the two helices forming the double helix are opposite hand and meeting at a common apex. They are produced by a variety of processes and are distinguished by the name of the manufacturing process by which the teeth are cut.

It is the usual practice when cutting this type of gear to know which direction of the teeth in relation to the driving end, especially in the case of shaft gears.

12.2.6   SPIRAL BEVEL GEARS

Spiral bevel gears (Figure 12.6) are identical to helical gears; the only reason they are called spiral gears is that the two gears are used to connect two shafts that are not parallel with each other.

Spiral gears used in this way give only a theoretical point contact instead of a line contact, as is the case for the previous gear types, and a longitudinal sliding motion is introduced between the teeth.

12.2.7   BEVEL GEARS

In the case of two intersecting shafts, bevel gears (Figure 12.7) would be used in this application. The most common type is that in which the teeth are radial to the point of intersection of the shaft axes or apex. Such gears are called straight bevel gears.

The teeth are similar to spur gears in that they make a line contact across the face of the teeth. The teeth are proportionally smaller at the front of the gear than the tooth form at the back of the gear in the ratio of:

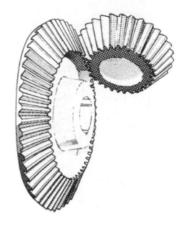

FIGURE 12.7   Bevel gears.

FIGURE 12.8   Spiral gears.

1. Distance from apex to the front end
2. Distance from the apex to the rear face

The shaft angle most commonly encountered is 90°, and the bevel gears with this angle are often referred to as miter gears.

12.2.8   SPIRAL GEARS

Spiral gears (Figure 12.8) are similar to helical gears; the only reason they are called spiral gears is that the two gears are used to connect two shafts that are not parallel with each other.

Spiral gears used in this way give only a theoretical point contact instead of a line contact, as is the case for the previous gear types, and a longitudinal sliding motion is introduced between the teeth.

12.2.9   WORM AND WORM WHEELS

Worm gears (Figure 12.9) are used in the same manner as spiral gears, i.e., to connect skew shafts, but not necessarily at right angles.

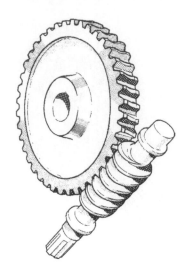

FIGURE 12.9   Worm and worm wheel.

The worm exactly resembles one of a pair of spiral gears, but the wheel is throated and has concave teeth. While resembling spiral gears in purpose and general dimensions, worm gears differ from them in that they give line contact instead of point contact and will carry greater loads. A further difference in spiral gears is that the center distance is important in relation to the helix angle. This problem does not arise when calculating the worm gears.

The sliding velocity is high when compared with other types of gears, and very special attention is necessary with the choice of materials and lubrication.

12.3   FORM OF TOOTH

The function of a pair of gears is to transmit motion from one shaft to another with a uniform velocity ratio. Either the cycloidal or the involute form of tooth will satisfy this requirement, but the cycloidal has the disadvantage that a slight variation in the theoretical center distance destroys the uniformity of angular velocity.

It is very difficult to ensure that the exact theoretical center distances can be maintained under all circumstances. The involute tooth form is much more tolerant to the small errors that can occur between the theoretical and manufactured center distances. This has ensured the universal adoption of the involute tooth form.

A further factor for the adoption of the involute tooth form is that it can be produced by a straight-sided rack cutter, and in fact every type of involute gear was originally produced by means of this basic principle. It can therefore be stated that all the types of gears produced in Section 12.2 are produced with an involute tooth form with the exception of bevel gears (Figure 12.7) and spiral bevel gears (Figure 12.8).

In the case of bevel gears, the teeth are generated by a straight-sided cutter, representing the side of a tooth of an imaginary crown gear, and the form of tooth produced is termed octoid.

A crown gear is in effect a bevel gear having a pitch angle of 90°, and the teeth of a true involute crown gear should have very slightly curved tooth profiles.

The difference between the tooth shapes on a gear being cut by an imaginary crown gear having a straight-sided cutter and one with a tooth profile that is slightly curved is so small that it can be ignored.

Both forms give theoretically correct action, and for practical purposes it can be safely assumed that the form of tooth produced on the transverse plane of both straight and spiral bevel gears is of involute form.

Spur gears (Figure 12.1), spur racks (Figure 12.3), helical gears (Figure 12.4), spiral gears (Figure 12.8), and worm and worm wheels (Figure 12.7) are all of involute form on their transverse planes, i.e., on the plane of rotation.

This point should be kept in mind when undertaking any calculations or layouts regarding the form of the tooth, especially in the case of helical gears, spiral gears, or worms.

For these types of gears, which in any case are closely related to each other, the form of tooth on the normal and axial planes is such that the form is entirely dependent on the involute form on the transverse plane and the spiral or helix angle.

12.4   LAYOUT OF INVOLUTE CURVES

The involute tooth form has been constructed on the transverse plane of the spur gear pair shown in Figure 12.1, and it will be particularly noted that the origin of the wheel and pinion tooth shapes is based on the base circle diameters D_o and d_o.

The base circle diameter should be considered as a disc around which a line of tape is wrapped around the diameter, the two ends of the tape commencing and finishing at point A, as shown in Figure 12.10.

By securing one end of the string or tape at position A, unwinding the free end of the tape will describe the path of the involute curve, through points B, C, D, and E, when the tape is tangential to the base disc at the corresponding points B_1, C_1 D_1, and E_1.

Before the involute form can be applied to a gear tooth, some considerations should be given to the question of the pitch diameter and pressure angle.

The pitch diameter and pressure angle together with the addendum and dedendum define which part of the involute curve is being utilized on a gear tooth, and before any tooth layout can made it is necessary to have this information.

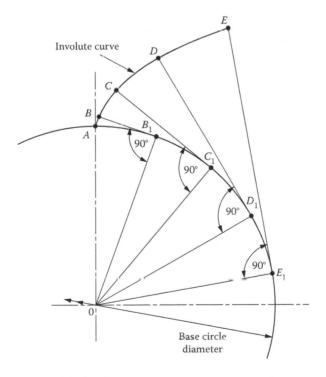

FIGURE 12.10   Development of the involute curve.

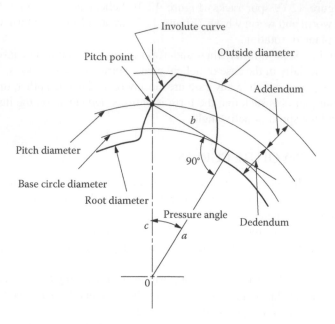

FIGURE 12.11 Description of the involute tooth.

The pitch diameter is always greater than the base circle, and where the involute bisects the pitch diameter circle, it is called the pitch point. From this point the pressure angle is obtained in conjunction with the base circle and the center of the gear.

In Figure 12.11 is shown the construction of a spur gear tooth in relation to the pressure angle and the pitch diameter; it should be noted that a right angle triangle is formed by the three sides a, b, and c when

$$a = \frac{\text{Base diameter}}{2} \tag{12.1}$$

$$b = \frac{\text{Pitch diameter}}{2} \tag{12.2}$$

$$\psi = \text{pressure angle of gear} \tag{12.3}$$

Therefore, given the pitch diameter D and the pressure angle ψ of a gear, the base circle diameter D_o is equal to D Cos ψ when D = 2xc and Do = 2xa.

Alternatively, given D and D_o, the pressure angle can be obtained as follows:

$$\text{Cos}\psi = \frac{D_o}{D} \tag{12.4}$$

There are several methods of setting out the involute tooth form, and the method selected is dependent on the accuracy required for the purpose for which it is proposed to use the tooth shape.

For example, if an accurate projection drawing is required, it is recommended that the tooth shape is generated, especially if a gear has a small number of teeth, as the shape changes very rapidly on gears of this nature.

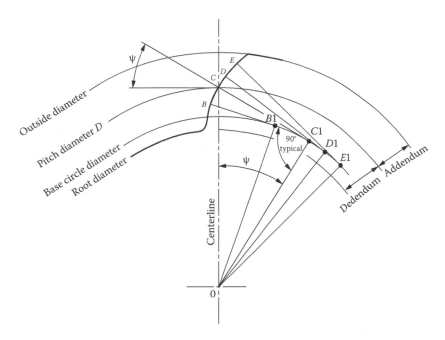

FIGURE 12.12 Layout of the involute tooth.

For a gear having a large number of teeth, the shape does not change, and in fact it will be found that the shape is practically equivalent to the arc of a circle.

For setting out tooth shapes when the accuracy is not the first essential, the following procedure can be followed. This method is suitable for helical, spiral, worm, and bevel gears on their transverse planes as well as spur gears.

Referring to Figure 12.12, the first step is to draw the pitch diameter D together with the outside and root diameters. The outside diameter is larger than the pitch circle diameter by an amount equal to twice the addendum. The root diameter is smaller than the pitch diameter by an amount equal to twice the dedendum of the gear. The pressure angle ψ is then set out from the center line of the gear and from the gear center O, while another line from the pitch point C at right angles to the line OC_1 is drawn to form a triangle OCC_1. The base circle is drawn equal to a radius of OC_1.

An arc is drawn from C to D measuring C_1C. A further arc is then drawn from D to E measuring D_1D. This process is repeated until the involute curve reaches the outside diameter and a similar procedure is adopted for the curve below the pitch diameter to the base circle.

In the example above, only four points have been used in the construction of the involute curve in the figure. More points could be used, and in fact greater accuracy will be obtained, but all this will be dependent on the size of the tooth being set out.

The other side of the tooth is drawn in an identical way after setting off the tooth thickness on the pitch line. The arcs used to construct the curve are exactly the same as were used for the first side.

12.5 INVOLUTE FUNCTIONS

In addition to being able to generate gear tooth profiles on their transverse planes in the manner described, it is also possible to plot the tooth form using coordinates previously calculated by means of involute trigonometry in conjunction with which involute functions are used.

To define an involute function, an involute tooth form is shown in Figure 12.13, where:

Ψ = Pressure angle of gear on transverse plane at radius D/2
D = Pitch circle diameter

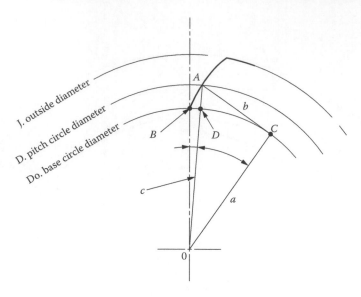

FIGURE 12.13   Involute function.

D_o = Base circle diameter
J   = Outside diameter
Ψ_1 = Involute function of angle ψ

The involute function is the vectorial angle (in radians) of an involute curve from its origin and is also a function of the pressure angle at the radius at which the vectorial angle is taken. The pressure angle at the point of origin B is zero.

The vectorial angle at A on the involute curve is therefore ψ_1, which is the angle subtended by OB and OA, the pressure angle at A at radius D/2 being ψ.

Therefore, the angle ψ_1 (in radians) is known as the involute function of the angle ψ and is written inv $\cdot \Psi$.

$$\text{Now } b/a = \tan \psi \text{ and } b = a \cdot \tan \psi$$

Length b is then rewound on the base circle diameter and is then equal to $a \cdot (\psi + \psi_1)$.
Therefore:

$$a \cdot (\psi + \psi_1) = a \cdot \tan \psi$$

$$(\psi + \psi_1) = \tan \psi$$

$$\Psi_1 = \tan \psi \qquad\qquad \Psi \text{ radians}$$

$$\Psi_1 = \text{inv} \cdot \psi$$

When the number of teeth are not an integer this will require one of the given variables to be modified, keeping as close to the theoretical values as possible. For spur gear dimensional terms see Figure 12.14, and for basic spur gear design formula, see Table 12.1.

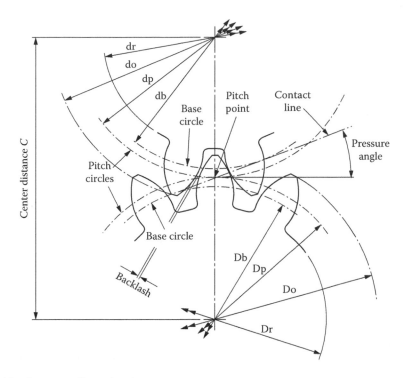

FIGURE 12.14   Spur gear dimensional terms.

TABLE 12.1
Spur Gear Design Formulae

Description		Formulae
Center distance	C	Given
Outside diameter	D_o	Given
Pressure angle	ψ	Given
Pitch diameter	D_p	$N \times m$
Number of teeth	N	D/m
Module	m	Given
Addendum	A	$1.00 \times m$
Dedendum	B	$1.25 \times m$
Circular pitch	P_c	$\pi D_p/N$, πm
Base circle diameter	Db	$Dp \times \cos(C)$
Root diameter	D_r	$Dp - (2.5 \times m)$

Example 12.1

To find the involute function of 20° the procedure is as follows:

Let: $\psi = 20° = 0.3490658$ radians

$$\text{Inv}\psi_1 = \tan\psi - \text{arc}\psi$$

where
$\tan\psi$ = natural tan of given angle
$\text{arc}\psi$ = numerical value of given angle in radians

Therefore:
$\text{Inv}\psi = \tan 20° - 20°$ (converted to radians)
$\text{Inv}\psi = 0.3639702 - (20 * 0.0174533)$

Note:
$1° = 0.0174533$ radians
$\text{Inv}\psi = 0.3639701 - 0.3490659$

Hence: $\text{inv}20° = 0.0149043$

 The involute function for 20° is 0.0149043, accurate to seven decimal places.
Using this procedure with the modern programmable calculators, it is safer to calculate the involute function rather than use the functions from a table of involute functions, as these may contain errors.

12.6   BASIC GEAR TRANSMISSION THEORY

Consider a simple gear drive with input and output shafts (Figure 12.15).
 In this simple case the gear ratio (n) is defined as

$$n = \frac{\text{input speed}}{\text{output speed}} = \frac{N_1}{N_2} \tag{12.5}$$

 Usually the speed of the shafts is described as rev/min, but the ratio will be the same if the units of angular velocity (ω) are used, i.e.,

$$n = \frac{\text{input speed}}{\text{output speed}} = \frac{\omega_1}{\omega_2}$$

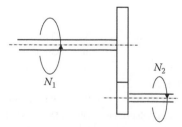

FIGURE 12.15   Simple gear drive.

12.6.1   TORQUE AND EFFICIENCY

The shaft power (SP) transmitted by a torque T Nm applied to a shaft that is rotating at N rev/min is given by

$$S_P = \frac{2\pi NT}{60}$$

In an ideal gear drive, the input and output powers are the same:

$$S_P = \frac{2\pi N_1 T_1}{60} = \frac{2\pi N_2 T_2}{60} \qquad N_1 T_1 = N_2 T_2$$

$$n = \frac{T_2}{T_1} = \frac{N_1}{N_2}$$

It follows then that as the input shaft speed is reduced, the torque will increase and vice versa. In a real drive or gearbox power is lost due to heat and friction and the power output will be reduced. The efficiency of the gear drive is defined by

$$\eta = \frac{\text{power out}}{\text{power in}} = \frac{2\pi N_2 T_2 \times 60}{2\pi N_1 T_1 \times 60} = \frac{N_2 T_2}{N_1 T_1} \tag{12.6}$$

As the input and output torques are different, the gearbox will require restraining to prevent it from rotating. A holding torque T_3 has to be applied to the gearbox through its attachments.

The total torque must equate to zero, i.e., $T_1 + T_2 + T_3 = 0$.

Using the convention that anticlockwise rotation is positive and clockwise is negative, the holding torque, together with its direction of rotation, can be determined. The direction of rotation of the output shaft will be dependent upon the internal configuration of the gears within the box.

12.7   TYPES OF GEAR TRAINS

12.7.1   SIMPLE GEAR TRAIN

Figure 12.16 shows a typical spur gears drive chain. The direction of rotation is reversed from one gear to another. The only function of the idler gear is to change the direction of rotation; it has no effect on the gear ratio.

t   = Number of teeth on the specific gear
D   = Pitch circle diameter (PCD).
m   = Modem = D/t
$D_A = m \cdot t_A$
$D_B = m \cdot t_B$
$D_C = m \cdot t_C$
ω = Angular velocity
v   = Linear velocity on the pitch circle diameter; $v = \omega.D/2$

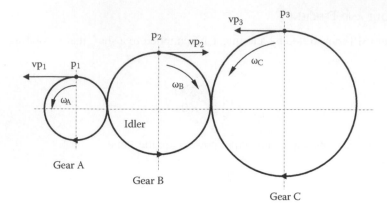

FIGURE 12.16   Simple gear train.

Considering the velocities of the gears at the pitch circles we have:

Linear velocity at point p_1 $$V_{p_1} = \frac{\omega_1}{D_1/2}$$ (12.7)

Linear velocity at point p_2 $$V_{p_2} = \frac{\omega_2}{D_2/2}$$ (12.8)

Linear velocity at point p_3 $$V_{p_3} = \frac{\omega_3}{D_3/2}$$ (12.9)

The velocity v of any point on the PCD must be the same for all the gears; otherwise they would be slipping. It follows that

$$\frac{\omega_A \cdot D_A}{2} = \frac{\omega_B \cdot D_B}{2} = \frac{\omega_C \cdot D_C}{2}$$

$$\omega_A \cdot D_A = \omega_B \cdot D_B = \omega_C \cdot D_C$$

$$\omega_A m t_A = \omega_B m t_B = \omega_C t_C$$ (12.10)

$$\omega_A t_A = \omega_B t_B = \omega_C t_C$$

$$N_A \cdot t_A = N_B \cdot t_B = N_C \cdot t_C$$

In terms of rev/min $$N_A \cdot t_A = N_B \cdot t_B = N_C \cdot t_C$$

If gear A is the input and gear C the output:

$$n = \frac{N_A}{N_C} = \frac{t_C}{t_A}$$ (12.11)

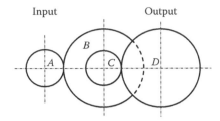

FIGURE 12.17 A compound gear set.

12.7.2 COMPOUND GEARS

A compound gear train has one or more gears fitted on one shaft, as illustrated in Figure 12.17.

In this example gears A and C are driving gears, and gears B and D are driven gears. Gears B and C are locked on the same shaft and therefore revolve at the same speed.

The velocity of each tooth on gears A and B is the same, so that $\omega_A.t_A = \omega_B.t_B$ as they are simple gears. Likewise for gears C and D, $\omega_C.t_C = \omega_D.t_D$.

$$\frac{\omega_A}{t_B} = \frac{\omega_B}{t_A} \quad \text{and} \quad \frac{\omega_C}{t_D} = \frac{\omega_D}{t_C}$$

$$\omega_A = \frac{t_B \cdot w_B}{t_A} \quad \omega_C = \frac{t_D \cdot \omega_D}{t_C}$$

$$\omega_A \cdot \omega_C = \frac{t_B \cdot \omega_B}{t_A} \times \frac{t_D \cdot \omega_D}{t_C} = \frac{t_B \cdot t_D}{t_A \cdot t_C} \times \omega_B \cdot \omega_D$$

$$\frac{\omega_A \cdot \omega_C}{\omega_B \cdot \omega_D} = \frac{t_B \cdot t_D}{t_A \cdot t_C}$$

As gears B and C are fitted on the same shaft, $\omega_B = \omega_C$

$$\frac{\omega_A}{\omega_D} = \frac{t_B \cdot t_D}{t_A \cdot t_C} = n$$

and since $\omega = 2\pi N$ then the gear ratio can be written as

$$\frac{N(in)}{N(out)} = \frac{t_B \cdot t_D}{t_A \cdot t_C} = n \tag{12.12}$$

12.8 POWER TRANSMISSION IN A GEAR TRAIN

In a gear train, power is lost between the teeth due to friction due to loads imposed on them and in the bearings. Power loss due to overcoming shaft inertia also contributes to the reduction in efficiency.

Consider the gear train for a hoist driven by an electric motor. The drive chain consists of two sets of reducing gears (Figure 12.18).

A motor is attached to the system with a moment of inertia I_m. The moment of inertia of the middle shaft is I_T, and I_H is the combined moment of inertia of the hoist; this acts as a load to the system. The gear ratio and gear efficiency of the gear set 1 – 2 are $n_{1/2}$ and $\eta_{1/2}$.

Gear set 3 – 4 consists of $n_{3/4}$ and $\eta_{3/4}$, respectively.

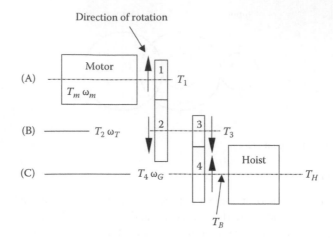

FIGURE 12.18 Diagram of motor/hoist gear drive.

Let: T_m = torque at the motor

T_H = torque of the hoist

T_B = friction torque of bearings

Draw a free body diagram using Newton's second law, $\Sigma T = I\alpha$.

In this instance, consider rotation in a clockwise direction as positive.

For shaft (A):

$$T_m - T_1 = I_m \cdot \alpha_m \qquad (12.13)$$

For shaft (B):

$$T_2 - T_3 = I_T \cdot \alpha_T \qquad (12.14)$$

Note: As there is a gear mating between gear 1 and gear 2, the analysis has to include its own gear ratio and gear efficiency and relate it to the transfer shaft, I_T.

Previously

$$\eta_{1/2} = \frac{T_2}{T_1} n_{1/2}$$

it follows that:

$$T_2 = \frac{\eta_{1/2} \cdot T_1}{n_{1/2}} \qquad (12.15)$$

For shaft (C):

$$T_4 - T_B - T_H = I_H \cdot \alpha_H$$

also
$$T_4 = \frac{T_3 \eta_{3/4}}{n_{3/4}}$$

(12.16)

Using power, $P = T_m . \omega_m$, the power transfer to each gear component is:

1. Power transfer by the motor:

$$P_M = T_m . \omega_m$$

2. Power at gear 1:

$$P_1 = T_1 \omega_m = (T_m - I_m . \alpha_m) \omega_m$$

3. Power at gear 2:

$$P_2 = P_1 . \eta_{1/2}$$

4. Power at gear 3:

$$P_3 = T_3 . \omega_T = (T_2 - I_T . \alpha_T) \omega_T$$

5. Power at gear 4:

$$P_4 = P_3 . \eta_{3/4}$$

6. Power at hoist:

$$P_H = T_H . \omega_H = (T_4 - T_B - I_H . \alpha_H) \omega_H$$

7. Overall power transfer efficiency, η_o:

$$\eta_o = \frac{P_H}{P_m}$$

Thus if friction torque, T_B effect is neglected.

$$\eta_o = \frac{P_H}{P_m} = \left| \frac{P_T}{P_m} \right| \times \left| \frac{P_H}{P_m} \right| = \eta_{1/2} \times \eta_{3/4}$$

Also;
$$\eta_o = \frac{T_H}{T_m} n_{1/2} \cdot n_{3/4}$$

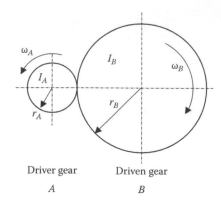

FIGURE 12.19   Referred inertia for a simple gear.

12.9   REFERRED MOMENT OF INERTIA, ($I_{REFERRED}$)

Consider a simple gear system. In order for the driver gear A to begin to rotate, it must have enough torque available to overcome its own inertia, IA, and then additional torque to begin to accelerate the driven gear B. The torque required to accelerate B needs to be initially calculated.

1. Torque at B to overcome I_B:

$$T_B = I_B.\alpha_B$$

   To refer α_B to gear A, the gear ratio is used:

$$n = \frac{\omega_B}{\omega_A} = \frac{\alpha_B}{\alpha_A}$$

   Thus $T_B = I_B \cdot n\alpha_A$.

2. The gear efficiency is related to power and torque of the mating gears:

$$\eta_G = \frac{P_B}{P_A} = \frac{T_B \cdot n}{T_A}$$

   where η_G is the efficiency of the gear set.

3. Torque required at gear A to accelerate I_B:

$$T_A = \frac{T_B \cdot n}{\eta_G} = \frac{I_B \alpha_B}{\eta_G} n = \frac{(I_B \cdot n\alpha_A)n}{\eta_G} = \frac{I_B n^2 \alpha_A}{\eta_G}$$

4. Total torque required at gear A to accelerate I_A and I_B:

$$T_{Total} = I_A \cdot \alpha_A + T_A$$

$$T_{Total} = \left[I_A + \frac{I_B n^2}{\eta_G} \right] \alpha_A$$

This can be expressed in a general form $T_{Total} = I_{equiv} \cdot \alpha_A$ when referred to gear A.

Hence

$$I_{equiv} = \left(I_A + \frac{I_B n^2}{\eta_G} \right)$$

This derivation of I_{equiv} for the simple gear system can be extended to cover for a double set of reducing gears, as discussed in Section 12.8, by neglecting the friction torque effect T_B.

Hence:

$$I_{equiv} = I_m + \frac{I_T \left(n_{1/2} \right)^2}{\eta_{1/2}} + \frac{I_G \left(n_{1/2} \right)^2 \left(n_{3/4} \right)^2}{\left(\eta_{1/2} \right) \left(\eta_{3/4} \right)}$$

12.10   GEAR TRAIN APPLICATIONS

Example 12.2

An electric motor accelerates a 500 kg load with an acceleration of 0.6 m/s² using a simple gear system (shown in Figure 12.20). The load is carried by a rope that encircles a hoist pulley that has a diameter of 1.0 m. The gear connected to the hoist has 200 teeth, and the gear connected to the motor shaft has 20 teeth. Assume the gear efficiency is 90%.

The masses and the radius of gyrations for each shaft are tabulated below:

	Mass (kg)	Radius of Gyration (mm)
Motor shaft	250	100
Hoist shaft	1,100	300

Calculate the torque required by the motor to raise the load with an acceleration of 0.6 m/s². Friction losses may be neglected.

Total torque required by motor to raise load:

$$T_{total} = T_m + T_{equiv}$$

where:

T_m = Torque to accelerate load through the gear system
T_{equiv} = Torque to overcome equivalent inertia (referred to the motor)

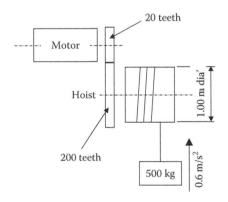

FIGURE 12.20   Diagram of hoist.

$$\text{From } I_{equiv} = I_m + \frac{I_H \cdot n^2}{\eta_g}$$

$$I_m = \text{motor shaft inertia}$$

$$I_m = m \cdot r^2$$

$$= 250 \, kg \times (0.1m)^2$$

$$= 2.5 \, kgm^2$$

$$I_H = 1100 \, kg \times 0.30 \, m$$

$$= 99.0 \, kgm^2$$

Motor shaft:

$$I_{equiv} = 2.5 \, kgm^2 + \left[\frac{99.0 \, kgm^2 \times 0.1^2}{0.9} \right]$$

$$= 3.6 \, kgm^2$$

Gear ratio (n):

$$n = \frac{N_1}{N_2}$$

$$= \frac{20}{200}$$

$$n = 0.10$$

Acceleration of hoist:

$$a_H = \alpha_H \cdot r_H$$

$$\alpha H = \frac{a_H}{r_H}$$

$$= \frac{0.6 \, m/s^2}{0.5 \, m}$$

$$= 1.2 \, rad/sec$$

From the gear ratio calculate the angular acceleration of the motor.

$$\alpha_m = \frac{\alpha_H}{n}$$

$$= \frac{1.2 \, rad/sec}{0.1}$$

$$= 12.0 \, rad/sec$$

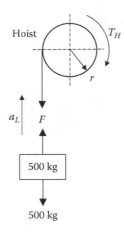

FIGURE 12.21 Torque required by hoist.

Torque due to equivalent inertia:

$$T_m = I_{equiv} \times \alpha_m$$

$$= 3.6 \times 12 \text{ rad/sec}$$

$$= 43.2 \text{ Nm}$$

Torque required at hoist (see Figure 12.21):

From Newton's second law:

$$\Sigma F = ma$$

$$F - 500g = 500a_H \quad F = 500 \, (g + a)$$

$$F = 500 \times (9.81 + 0.6)$$

$$F = 5,205 \text{ N}$$

Torque at hoist:

$$T_H = F \cdot r$$

$$= 5,205 \text{ N} \times 0.5 \text{ m}$$

$$= 2,602 \text{ Nm}$$

Due to the gear efficiency (as the hoist is connected to the gear system):

Torque required to accelerate the load:

$$T_{equiv} = \frac{T_H \cdot n_{1/2}}{\eta_{1/2}}$$

$$= \frac{2602.5 \times 0.1}{0.9}$$

$$= 289.17 \text{ Nm}$$

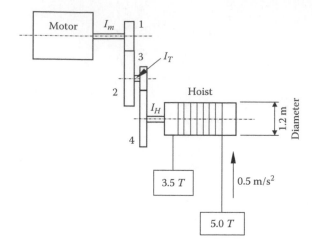

FIGURE 12.22   Diagram of hoist mechanism for Example 12.3.

Total torque referred to motor:

$$T_{total} = 43.2 + 289.17 \text{ Nm}$$

$$= 332.37 \text{ Nm}$$

Example 12.3

An electric motor is used to raise and accelerate a load using a hoist and driving through two sets of reducing gears as in Figure 12.22.

	Moment of Inertia kgm²	Gear Ratio
Motor shaft, I_m	5	Set ½ = 1:3.5
Transfer shaft, I_T	40	
Hoist shaft, I_H	500	Set ¾ = 1:4.5

Gear efficiency for both sets of gears is 90%. Neglecting friction, determine the total torque required by the motor to accelerate a load of 5 tonnes at an acceleration of 0.5 m/s².
   Given that:

$$I_m = 5 \, \text{kgm}^2$$

$$I_T = 40 \, \text{kgm}^2$$

$$I_H = 500 \, \text{kgm}^2$$

$$n_{1/2} = 1 : 3.5$$

$$n_{3/4} = 1 : 4.5$$

$$\eta G = 0.9$$

Neglecting friction effects:

Total torque required by the motor is

$$T_{total} = T_{m1} + T_{m2}$$

where:

T_{m1} = Torque to overcome equivalent inertia (referred to motor)
T_{m2} = Torque to accelerate the load through the gear system

Consider T_{m1}:

$$T_{m1} = I_{equiv}\,\alpha_m$$

For a compound gear set neglecting friction:

$$I_{equiv} = I_m + \frac{I_T \cdot (n_{1/2})^2}{\eta_{G1/2}} + \frac{I_H \cdot (n_{1/2})^2 (n_{3/4})}{(\eta_{G1/2}) \cdot (\eta_{G3/4})}$$

$$I_{equiv} = 5 + \frac{40(1/3.5)^2}{0.9} + \frac{500(1/3.5)^2 (1/4.5)^2}{(0.9) \cdot (0.9)}$$

$$= 11.116\,Nm$$

Given that the linear acceleration of the load $a_H = 0.5$ m/s²,

$$\alpha_H = \frac{a_H}{r_H}$$

$$= \frac{0.5\ m/s^2}{0.5\ m}$$

$$= 1.0\ rad/s^2$$

From the gear ratio:

$$\frac{\alpha_H}{\alpha_m} = \frac{\alpha_H}{\alpha_T} \times \frac{\alpha_T}{\alpha_m}$$

$$= \frac{1}{4.5} \times \frac{1}{3.5}$$

$$= \frac{1}{15.75}$$

Thus,

$$\alpha_m = 15.75 \cdot \alpha_H$$

$$\alpha_m = 15.75 \cdot (1.0)$$

$$= 15.75\ rad/s^2$$

Therefore,

$$T_{m1} = I_{equiv} \cdot \alpha_m$$
$$= 11.116 \times 15.75$$
$$= 175.08 \text{ Nm}$$

Determine T_{m2}:

From Newton's second law $\Sigma F = ma$

$$3500 - F_1 = 3500 \cdot a$$
$$F_1 = 3500 (g - a)$$
$$= 3500 (9.31)$$
$$= 32585 \text{ N}$$

$$F_2 - 5000 \text{ g} = 5000 \cdot a$$
$$F_2 = 5000 (g - a)$$
$$= 46550 \text{ N}$$

Resultant force at hoist:

$$F_r = F_2 - F_1$$
$$= 13,965 \text{ N}$$

Torque at hoist:

$$T_H = F_r \times r_H$$
$$= 13,965 \times 0.5$$
$$= 6,982.5 \text{ Nm}$$

$T_{m2} = T_H$ referred to the motor.

$$T_{m2} = \frac{T_H \ n_{1/2} n_{3/4}}{\eta_{1/2} \cdot \eta_{3/4}}$$

$$T_{m2} = \frac{6982.5 (1/3.5)(1/4.5)}{(0.9)(0.9)}$$

$$= 547.33 \text{ Nm}$$

Total torque at the motor is

$$T_{total} = T_{m1} + T_{m2}$$
$$= 175.08 + 547.33$$
$$T_{total} = 722.41 \text{ Nm}$$

13 Introduction to Cams and Followers

13.1 INTRODUCTION

Modern-day automated production machinery relies on servo-driven motion control systems, and this has unquestionably provided considerable flexibility in the design of machine units. This has led to a revolution in machine design where motion control can be accomplished using servo-motors that are connected to its control system by cables instead of drives and linkages.

An added advantage is that the "cam profile" is stored electronically and can be amended within the control unit program without the need to modify or machine a new cam profile.

It is important for the design engineer to be aware of the history of the early motion control and the reasons for its evolution.

Cams are a very versatile mechanism with many applications. They generate motion, including rectilinear, oscillatory, or rotary motion to a mechanical element nominally referred to as a follower. This is in contact with the cam profile.

Many types of motion are possible for the follower, as the cam face can have any desired profile within certain limitations, including acceptable dynamic forces.

The cam is generally attached to a rotating shaft. In some cases the rotation may not be continuous but can be oscillatory. The cam profile will determine the type of motion the follower will have, such as a slow rise at a particular acceleration to a given value followed by a dwell for a limited period and a rapid return to the start point of the cycle.

13.2 BACKGROUND

In the early and mid twentieth century, high production automatic machinery was exclusively driven by cams and gears. In the late twentieth century the electric servo and hydraulic servo systems began to be developed with a high degree of reliability and offered a more versatile operating system. Although this was thought to be the death knell of the use of cams in machinery, this has not occurred across the board. There are numerous examples, including packaging machinery, labeling, etc., where repetitious and accurate movements are required to position or provide a stop within a process line. These are usually provided by a cam-operated mechanism, as it generally offers a cost-effective and reliable method. Most people's thoughts generally turn to the automotive engine, where cams are used to operate the valve gear for the induction and exhaust process. This is undertaken under the most adverse conditions of heat and vibration and provides a very reliable and cheap system that has not found any alternative to replace it.

This section will provide the design engineer with some advice on the design of a suitable cam-operated system using either a plate cam or a cylindrical cam. An imaginary project is considered such that the various elements can be demonstrated and the effect on the overall design is considered.

13.3   REQUIREMENTS OF A CAM MECHANISM

The basic requirements for a cam-operated mechanism should consider the following points:

1. The mechanism should be "stiff" and have minimal deflections in any linkages, etc.
2. The cam follower must be able to maintain contact with the cam profile under all operating conditions.
3. The mechanism must be able to operate and maintain positional tolerances.
4. The mechanism must be able to operate at the design speed and maintain clearances with other mechanisms operating within the same space envelope.
5. The output from the follower must meet the design requirements.

13.4   TERMINOLOGY

There are two principal designs of rotary cams: plate cams and cylindrical cams.

13.4.1   PLATE CAMS

Plate-type cams are the most common and recognizable form of a cam, where examples can be found in automotive engines for the operation of the inlet and exhaust valves where the cam is machined onto the "cam shaft." Other examples will be found in automatic machine tools, packaging machinery, and some quick-release work vices. In general, the profiles can easily be machined using standard machine tools.

The cam followers for plate-type cams come in two forms: sliding follower and rotary or swinging arm follower.

The cam followers move in a plane perpendicular to the cam face.

Figures 13.1 and 13.2 show both forms of follower and also show the terminology used in the cam designations.

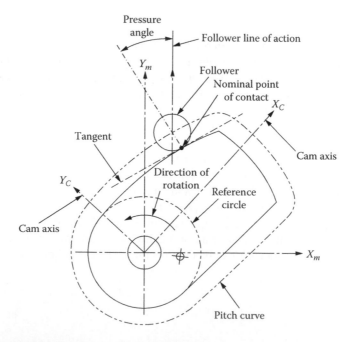

FIGURE 13.1   Cam driving an offset translating follower.

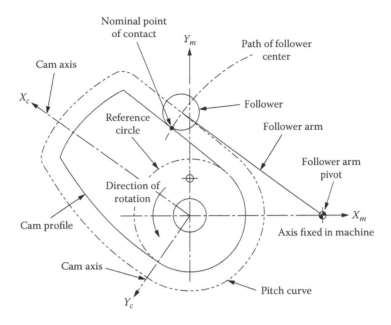

FIGURE 13.2 Cam driving a swinging arm follower.

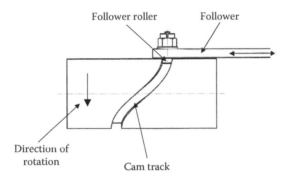

FIGURE 13.3 Cylindrical cam and follower.

13.4.2 CYLINDRICAL CAMS

In certain instances it may not be possible to fit a plate-type cam due to space limitations. In this instance a cylindrical cam (Figure 13.3) may be a more convenient solution.

In this design the cam follower moves in a plane parallel to the axis of the cam.

Depending on the size of the cam, the cam track may be made up of segments allowing for the track to be modified to meet particular requirements. This type of cam and follower arrangement is more likely to be found in earlier versions of metal-cutting automatic lathes where the follower connects to a cross-slide carrying the cutting tool. The track will advance the follower at a specific linear rate for either form cutting or even cross-drilling, and on completion this will be followed by a short dwell and then a rapid return to its starting position.

An advantage to the use of the cylindrical cam is the roller follower is essentially trapped in the cam track and therefore follows the cam profile accurately. The mechanism is sometimes fitted with a light spring to keep the roller against one side of the track when returning to the start position.

A further variation to the cylindrical cam is where the cam profile is machined onto the edge of the cylinder. These are referred to as a face or shell cams.

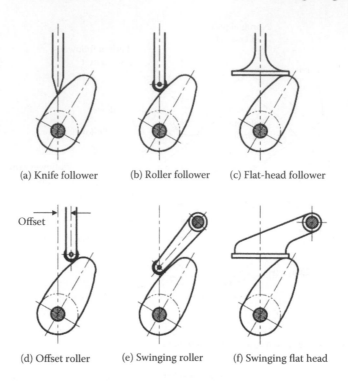

(a) Knife follower   (b) Roller follower   (c) Flat-head follower

(d) Offset roller   (e) Swinging roller   (f) Swinging flat head

FIGURE 13.4   Typical cam follower designs.

13.4.3   Typical Cam Follower Arrangements for Plate-Type Cams

Figure 13.4 illustrates a selection of cam followers commonly used on plate-type cams:

(a) Depicts a knife edge follower. These are seldom used, as the stresses at the tip of the follower are very high and subject to rapid wear due to sliding at the cam face. They are only used in slow-moving cams.
(b) Roller followers are more commonly used as they eliminate any sliding between the roller and the cam face. They tend to be used in moderate to high-speed systems.
(c) Although friction between the cam face and the flat head followers may present a problem without correct surface treatment of the follower and cam face, the advantage of this type of follower is that it eliminates any side thrust between the contact surfaces.

Figure 13.4(d)–(f) shows variations from (b) and (c), with the centerline position of the roller position in (d) being offset, and (e) and (f) show swinging cam followers for the roller and flathead designs.

The advantage of using this type of cam followers in (e) and (f) is the pressure angle of the cam system can be increased. This will be discussed in Section 13.7.

13.5   THE TIMING DIAGRAM

The timing diagram is an important feature in the cam design. As the name suggests, this covers all aspects of the mechanism's operating envelope, including the cam rise time, dwell, return, and final dwell, usually the base circle of the cam.

Once a preliminary design has been selected, consideration should be given to its operation using the timing diagram (Figure 13.5). This will investigate the feasibility of the design and highlight any potential problems, including any possible crash points in the design.

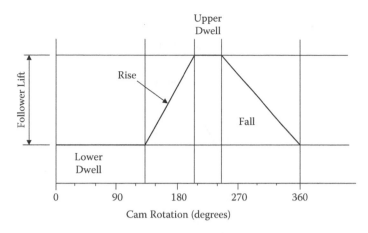

FIGURE 13.5 Typical timing diagram.

The timing diagram is only concerned with the angular rotation of the cam or the timescale the cam operation is based upon. The shape of the cam profile is irrelevant, as the timing diagram is only concerned with the start of the lift period and its completion. Other machine activities may be in operation at the same time, and the diagram will help to synchronize these operations.

13.6 CAM LAWS

The cam profile is important as it dictates the following:

1. The speed of rotation of the cam
2. How the follower behaves (will the follower leave the profile?)
3. The life of the follower

The profile can be classified by examination of the follower motion requirements, investigating the velocity, acceleration, and jerk generated due to the profile.

The cam laws cover and include the following:

Constant velocity
Parabolic (constant) acceleration
Simple harmonic motion
Cycloidal (sine) acceleration

The first step is to identify the cam rotation required to lift (or drop) the follower from the lower dwell to the upper dwell point on the cam profile. This is best carried out using the timing diagram (Figure 13.6). In constructing the diagram, a straight line is drawn between the lower and upper points between the dwell points, without giving consideration to the profile requirements at this point.

The cam duration can be expressed in either degrees (θ) or time in seconds (t).

The following expression can describe the follower displacement y due to a rotation θ of the cam:

$$y = f(\theta) \quad \text{or} \quad y = f(t) \tag{13.1}$$

depending on the units of the x axis used.

The first differential of y with respect to t gives the velocity of the follower:

$$v = \frac{dy}{dt} \quad \dot{y} = \frac{df(t)}{dt} = \frac{df(\theta)}{d\theta}\frac{d\theta}{dt} = \omega\frac{df}{d\theta} \tag{13.2}$$

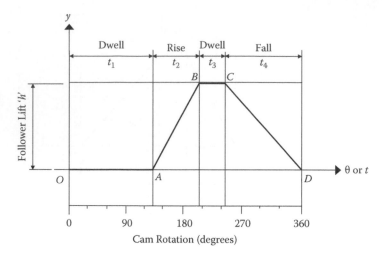

FIGURE 13.6 Timing diagram.

where ω = angular velocity of the cam.

The second differential of y with respect to t will give the acceleration of the follower:

$$a = \frac{d^2y}{dt} \quad \ddot{y} = \omega^2 \frac{d^2f}{dq^2} + \dot{\omega} \frac{df}{dq} \tag{13.3}$$

The third differential of y with respect to t will give the jerk* of the follower:

$$J = \frac{d^3y}{dt^3} = \dddot{y} = \omega \frac{df}{dq} + 3\omega\dot{\omega} \frac{d^2f}{dq^2} + \omega^3 \frac{d^3f}{dq^3}$$

In the case of the cam rotating at a constant rotational speed the above equations for acceleration can be reduced to the following:

Acceleration:

$$a = \omega^2 \frac{d^2f}{d\theta^2}$$

Jerk:

$$J = \omega^3 \frac{d^3f}{d\theta^3}$$

let   T = time taken for the cam to ratate through an angle of α degrees

$$= \frac{\alpha}{\omega}$$

where α is the rotation of the cam in degrees for a follower rise of h.

If a cam is designed to have a rise or fall at a constant velocity, it will experience very large changes in acceleration and jerk.

---

* Jerk is an undesirable feature, as it can result in the follower leaving the cam profile momentarily, or at its worst, resulting in substantial damage to the follower.

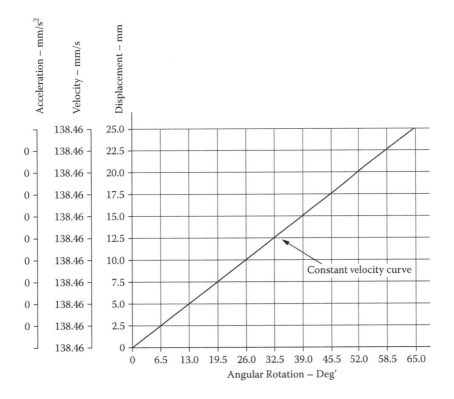

FIGURE 13.7   Constant velocity curve.

13.6.1   CONSTANT VELOCITY OF THE FOLLOWER

In this first case, the follower moves at a constant velocity from the base circle O to the maximum lift of the follower A (Figure 13.7).

Consider a nominal cam profile with the following characteristics:

h = 25.0 mm   Full lift
α = 65°   Angular rotation to full lift
θ = 6.5°   Interval spacing
ω = 360°/s   Angular velocity

Displacement:
$$y = \frac{ht}{T} \quad \text{or} \quad y = \frac{h\theta}{\alpha}$$

Velocity:
$$v = \frac{h}{T} \quad \text{or} \quad v = \frac{h\omega}{\alpha}$$

Acceleration:
$$a = 0$$

At positions O and A where the acceleration will be infinite. Tabulating the above using a set of notional values.

Table 13.1 shows the above using a set of notional values.

This type of motion is unacceptable except at very low speeds.

TABLE 13.1

Constant Velocity Values

y = 25 mm
α = 65°
θ = 2.5°
w = 360°/s

y mm	θ°	v mm/s	a mm/s²
0.0	0.0	138.46	Infinity
2.5	6.5	138.46	0
5.0	13.0	138.46	0
7.5	19.5	138.46	0
10.0	26.0	138.46	0
12.5	32.5	138.46	0
15.0	39.0	138.46	0
17.5	45.5	138.46	0
20.0	52.0	138.46	0
22.5	58.5	138.46	0
25.0	65.0	138.46	Infinity

13.6.2 PARABOLIC MOTION

In this second case (Figure 13.8), the profile is also known as a constant acceleration cam law. This law has been widely used in the past, as it produces the lowest maximum follower acceleration. It does produce an infinite jerk at both the beginning and the end of the motion cycle and also at the point of inflection. This makes this a very poor choice where dynamic considerations are important, and for this reason it is only suitable for low-speed systems.

Mathematically we have:

Displacement:
$$y = 2h\left(\frac{t}{T}\right)^2 \quad \text{or} \quad y = 2h\left(\frac{\theta}{\alpha}\right)^2$$

Velocity:
$$v = 4h\frac{t}{T2} \quad \text{or} \quad v = \frac{4h\omega\theta}{\alpha^2}$$

Acceleration:
$$a = \frac{4h}{T} \quad \text{or} \quad a = \frac{4h\omega^2}{\alpha^2}$$

Jerk: $J = 0$ Except at o, B and A where it is infinite.

The tabulated values of the notional cam are shown in Table 13.2.

Due to the sharp increase and decrease in the acceleration curve, jerk will be extremely high, resulting in the possibility of the follower leaving the cam profile. For this reason this kind of profile is only suitable for low-speed systems.

13.6.3 SIMPLE HARMONIC MOTION

This third case of cam laws shows an improvement on the velocity and acceleration curves compared that of the parabolic cam law in that they are smoother without any discontinuities.

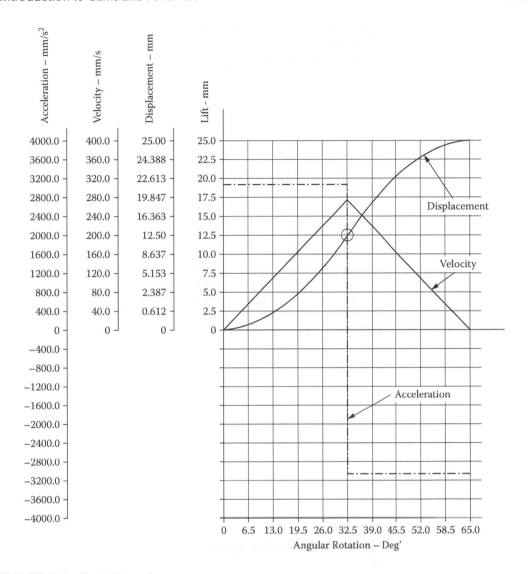

FIGURE 13.8 Parabolic motion.

See Table 13.3 to consider the mathematics and Figure 13.9 to compare the displacement, velocity, acceleration, and jerk curves for this type of motion.

Displacement:
$$y = \frac{h}{2}\left(1 - \cos\frac{\pi t}{T}\right) \quad \text{or} \quad y = \frac{h}{2}\left(1 - \cos\frac{\pi\theta}{\alpha}\right)$$

Velocity:
$$v = \frac{\pi h}{2T}\sin\frac{\pi t}{T} \quad \text{or} \quad v = \frac{\pi h\omega}{2\alpha}\sin\frac{\pi\theta}{\alpha}$$

Acceleration:
$$a = \frac{\pi^2 h}{2T^2}\cos\frac{\pi t}{T} \quad \text{or} \quad a = \frac{\pi^2 h\omega^2}{2\alpha^2}\cos\frac{\pi\theta}{\alpha}$$

Jerk:
$$J = -\frac{\pi^3 h}{2T^3}\sin\frac{\pi t}{T} \quad \text{or} \quad J = -\frac{\pi^3 h\omega^3}{2\alpha^3}\sin\frac{\pi\theta}{\alpha}$$

TABLE 13.2
Parabolic Motion Values

y = 25 mm
α = 65°
θ = 2.5°
ω = 360°/s

Y mm	θ°	Disp. mm	v mm/s	a mm/s²	J mm/s³
0.0	0.0	0.0	0.0	3,067.46	0
2.5	6.5	0.5	55.4	3,067.46	0
5.0	13.0	2.0	110.8	3,067.46	0
7.5	19.5	4.5	166.2	3,067.46	0
10.0	26.0	8.0	221.5	3,067.46	0
12.5	32.5	12.5	276.9	0.00	0
15.0	39.0	17.0	221.5	−3,067.46	0
17.5	45.5	20.5	166.2	−3,067.46	0
20.0	52.0	23.0	110.8	−3,067.46	0
22.5	58.5	24.5	55.4	−3,067.46	0
25.0	65.0	25.0	0.0	−3,067.46	0

TABLE 13.3
Simple Harmonic Motion Values

y = 25 mm
α = 65°
θ = 2.5°
ω = 360°/s

y mm	θ°	Disp mm	v mm/s	a mm/s²	Jerk mm/s³
0.0	0.0	0.000	0.00	3,784.32	0.0
2.5	6.5	0.612	67.21	3,599.10	−20,347.4
5.0	13.0	2.387	127.84	3,061.58	−38,703.1
7.5	19.5	5.153	175.96	2,224.37	−53,270.2
10.0	26.0	8.637	206.85	1,169.42	−62,622.9
12.5	32.5	12.500	217.49	0.00	−65,845.6
15.0	39.0	16.363	206.85	−1,169.42	−62,622.9
17.5	45.5	19.847	175.96	−2,224.37	−53,270.2
20.0	52.0	22.613	127.84	−3,061.58	−38,703.1
22.5	58.5	24.388	67.21	−3,599.10	−20,347.4
25.0	65.0	25.000	0.00	−3,784.32	0.0

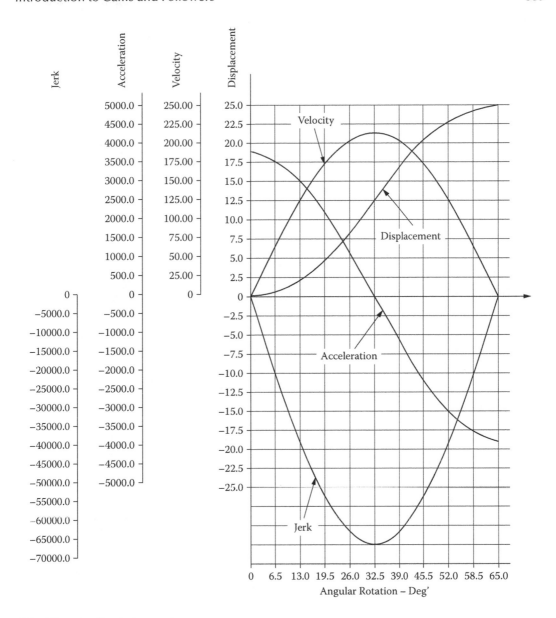

FIGURE 13.9 Simple harmonic curve.

13.6.4 CYCLOIDAL MOTION

This fourth case (Figure 13.10) has the highest nominal acceleration of the cam laws reviewed so far. This case also has a finite jerk.

Although the acceleration has a greater value than that of the simple harmonic motion for the same cam parameters, this curve is widely used in high-speed applications. Since the curve has zero acceleration at the beginning and the end of the motion segment, it is easily coupled to dwells at these points.

Coupling two cycloidal curves together without a dwell period in between is to be avoided, as the pressure angle will tend to be high in this instance. See Table 13.4 for the tabulated values.

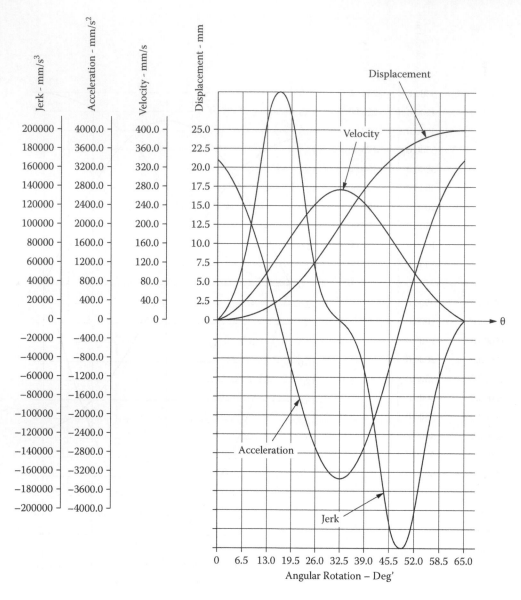

FIGURE 13.10   Cycloidal motion curve.

The mathematics for the cycloidal curve:

Displacement: $$y = h\left(\frac{t}{T} - \frac{1}{2\pi}\sin\frac{2\pi t}{T}\right) \quad \text{or} \quad y = h\left(\frac{\theta}{\alpha} - \frac{1}{2\pi}\sin\frac{2\pi\theta}{\alpha}\right)$$

Velocity: $$v = \frac{h}{T}\left(1 - \cos\frac{2\pi t}{T}\right) \quad \text{or} \quad v = \frac{h\omega}{\alpha}\left(1 - \cos\frac{2\pi\theta}{\alpha}\right)$$

Acceleration: $$a = \frac{2\pi h}{T^2}\sin\frac{2\pi t}{T} \quad \text{or} \quad a = \frac{2\pi h\omega^2}{\alpha^2}\sin\frac{2\pi\theta}{\alpha}$$

Jerk: $$J = -\frac{4^3 h}{T^3}\cos\frac{2\pi t}{T} \quad \text{or} \quad J = -\frac{4\pi^2 h\omega^3}{\alpha^3}\cos\frac{2\pi\theta}{\alpha}$$

TABLE 13.4

Cycloidal Motion Values

$y = 25$ mm

$\alpha = 65°$

$\theta = 2.5°$

$\omega = 360°/s$

y mm	$\theta°$	Disp mm	v mm/s	a mm/s^2	J mm/s^3
0.0	0.0	0.0	0.0	0.00	167,675
2.5	6.5	0.2	26.4	2,832.15	135,652
3.5	13.0	1.2	95.7	4,582.52	51,814
7.5	19.5	3.7	181.2	4,582.52	−51,814
10.0	26.0	7.7	250.5	2,832.15	−135,652
12.5	32.5	12.5	276.9	0.00	−167,675
15.0	39.0	17.3	250.5	−2,832.15	−135,652
16.0	45.5	21.3	181.2	−4,582.52	−51,814
20.0	52.0	23.8	95.7	−4,582.52	51,814
22.5	58.5	24.8	26.4	−2,832.15	135,652
25.0	65.0	25.0	0.0	0.00	167,675

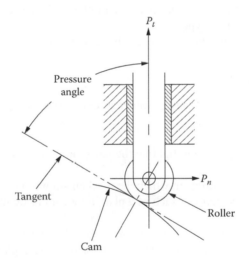

FIGURE 13.11   Pressure angle.

13.7   PRESSURE ANGLE

The pressure angle at any point on the cam profile can be defined as the angle between the follower direction of motion and the normal to the tangent at the point of contact of the follower with the cam profile (see Figure 13.11). The value of the pressure angle is important, as it dictates the forces acting on the follower. From Figure 13.11 it can be seen that the force acting at the pressure angle α can be resolved into two components, one tangential to the direction of travel and the second at right angles to it. This angle α is referred to as the pressure angle. Therefore,

$$P = P_t + P_n$$

P_n is undesirable as it exerts a side thrust on the follower guides or bearings, and if this is large enough, it will cause the follower to bind and increase the friction.

It is common practice for conventional translating-type followers to keep the pressure angle below 30° and at or below 45° for swinging-type followers. These values are on the conservative side and in many cases may be increased, but beyond these limits trouble could develop and further analysis will be required.

The pressure angle is important for the following reasons:

- As stated above, the pressure angle increases the side thrust acting on the follower, and this in turn increases the forces between the follower and cam profile, resulting in undesirable wear.
- Reducing the pressure angle increases the size of the cam, and often this is undesirable where space is at a premium. Furthermore, large cams require greater care in manufacture.
- Larger cams also mean more mass, and hence higher inertia forces. This in turn will lead to increased vibration in high-speed machines.
- The inertia of the larger cam may impede with quick starting and stopping.

There are ways to reduce the pressure angle; these include:

- Increase the base circle diameter.
- Reduce the total follower travel.
- Increase the amount of cam rotation for a given follower displacement.
- Change the follower motion type.
- Change the follower offset; this may generally be reduced.

13.8   DESIGN PROCEDURE

The following is a recommended design procedure to establish an optimum cam-operated mechanism.

1. Determine a timing diagram highlighting the key requirements for displacements, velocities, and accelerations.
2. Determine the space constraints in which the mechanism will be required to fit.
3. Select a suitable cam mechanism such as a plate-type cam or cylindrical cam that will fit best in the available space.
4. Consider the cam laws the mechanism will be required to meet.
5. Calculate the minimum size of cam that will meet the pressure angle and profile constraints.
6. Calculate the maximum contact stress and dynamic load on the follower.
7. Do the dimensions of the cam mechanism and materials meet the design requirements?
8. Will an external supplier manufacture the cam mechanism? If so, ensure it is in the consultation loop.
9. Finally, prepare a product design specification (PDS).

13.9   GRAPHICAL CONSTRUCTION OF A CAM PROFILE

In the graphical design of cams the problem of construction is simplified by employing inversion, that is, to imagine the cam being held stationary and the follower then being rotated in the opposite direction to the cam rotation. This preserves the correct sequence when the motion takes place.

To illustrate the point, Figures 13.12 and 13.13 are shown; the profile of the cam and the follower is then placed in equal increments around the periphery.

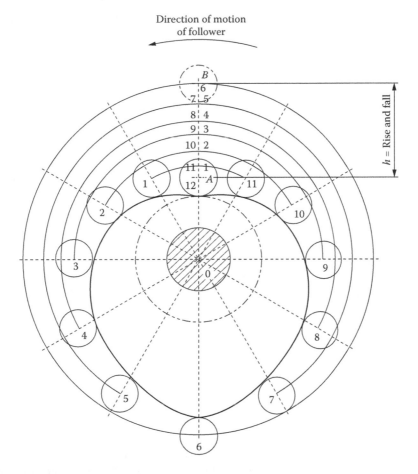

FIGURE 13.12 Constant velocity of follower–graphical construction.

The first construction (Figure 13.12) shows a cam designed for imparting a constant velocity on a translating-type roller that is on the centerline of the cam. For the purposes of the construction, one complete revolution of the cam is to give one rise and one fall to the follower.

Example 13.1

A plate cam fitted with a translating-type follower (Figure 13.12) with the following details is to be designed:

Shaft diameter = 10.0 mm
Base circle diameter = 23.0 mm
Roller diameter = 7.0 mm
Rise and fall of follower = 15.0 mm
Rotation of the cam = clockwise

Step 1. Draw the shaft diameter and base circle at center O and extend the centerline upwards to represent the follower position. The original drawing was produced using AutoCad; therefore it was drawn full size.

Step 2. Offset the base circle diameter plus half the radius of the follower diameter by the rise dimension (15.0 mm) to position B. Considering the cam is held stationary, the follower will rotate in a counterclockwise direction.

Step 3. Subdivide the line A:B into six equal parts, numbering them 1, 2, 3, ..., and produce a
series of arcs continuing around the circle as shown in Figure 13.12.
Step 4. Radiate from the center O a series of lines at 30° angular rotation.
Step 5. Locate positions 1, 2, 3, ..., and on the intersections draw the rollers.
Step 6. If you are using a drafting program such as AutoCad, the next step is to draw straight
lines between the points of intersection, following this with a "PEDIT" command. The lines
are joined and followed by the "SPLINE-FIT." This will generate a suitable curve.

Example 13.2

It is required to design a cam using a swinging link that is fitted with a roller follower. The profile
is to have uniform velocity for both the rise and fall segments of the cam.
Cam data:

Shaft diameter = 10.0 mm
Base circle diameter = 23.0 mm
Roller diameter = 7.0 mm
Position of the pivot from the cam centerline:
    X = 30.0 mm
    Y = 15.0 mm
Total angular movement of the link = 30°
Rotation of the cam = clockwise

Refer to Figure 13.13.

Step 1. Draw the shaft diameter and base circle at center O and position the swinging link at
position P with a length P:A. The original drawing was produced in AutoCad; therefore, it
was drawn full size.
Step 2. Draw a circle with a radius of O:P in order for the successive positions of the link to be
positioned. Subdivide this circle into 12 equal parts and number them as P1, P2, P3, ..., P12.
Considering the cam is held stationary, the follower will rotate in a counterclockwise direction.
Step 3. From the position P construct an arc of a circle A:B with a radius P:A such that the angle
A:P:B = 30°. Divide this arc into six equal parts, numbering them 1, 2, 3, ..., and produce a
series of arcs continuing around the circle.
Step 4. From each position of P1, P2, P3, ..., construct an arc as in step 2, originating from P2,
P3, etc.
Step 5. Locate points 1, 2, 3, ..., on these lines and with these points as centers; then draw
the rollers.
Step 6. As in the previous example, if you are using AutoCad, "PEDIT" and "SPLINE-FIT" will
produce a suitable curve.

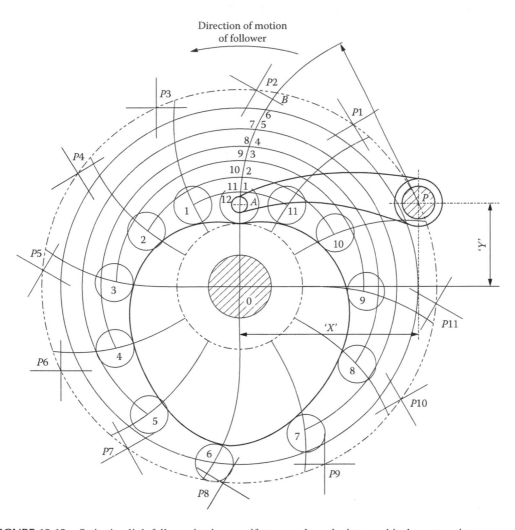

FIGURE 13.13 Swinging link follower having a uniform angular velocity–graphical construction.

Index